The Physical Microbe

An introduction to noise, control, and communication in the prokaryotic cell

The Physical Microbe

An introduction to noise, control, and communication in the prokaryotic cell

Stephen J Hagen

Physics Department, University of Florida

Morgan & Claypool Publishers

Rights & Permissions
To obtain permission to re-use copyrighted material from Morgan & Claypool Publishers, please contact info@morganclaypool.com.

ISBN   978-1-6817-4529-9 (ebook)
ISBN   978-1-6817-4528-2 (print)
ISBN   978-1-6817-4531-2 (mobi)

DOI   10.1088/978-1-6817-4529-9

Version: 20171001

IOP Concise Physics
ISSN 2053-2571 (online)
ISSN 2054-7307 (print)

A Morgan & Claypool publication as part of IOP Concise Physics
Published by Morgan & Claypool Publishers, 1210 Fifth Avenue, Suite 250, San Rafael, CA, 94901, USA

IOP Publishing, Temple Circus, Temple Way, Bristol BS1 6HG, UK

For Melanie, Sam and Noah.

Contents

Preface: Why the physical microbe?

There are no small parts, only small actors.

Constantin Stanislavski

The microbial cell is a sophisticated machine. It builds and maintains itself using an elaborate molecular toolbox that it regulates with precision. It imports and processes nutrients, senses its environment, communicates with its cohort, and turns information into action. It grows and multiplies with alacrity. Leafing through any microbiology text, one cannot help being struck by the difficult technical and physical problems a micron-sized cell has solved in order to thrive: making decisions using chemical circuits, preparing optimally for unpredictable environments, communicating efficiently with competing or cooperating organisms, and growing as rapidly as possible when conditions are favorable. It seems one could reinterpret the standard microbiology text as an overview of the physical problems that a unicellular organism must solve, together with real world examples of how to solve them. In fact there are some excellent biophysics or physical biology textbooks that take this approach, disassembling large parts of biology and then reassembling the subject from a physical perspective.

The goal here is less ambitious. This book does not attempt any kind of comprehensive overview of cell biology, microbiology, or even biophysics. It is not even an introduction to microbiology. Instead it aims to use microbes to introduce a few major themes in physical biology. These themes include gene regulation, networks, noise and heterogeneity, and signaling and information flow. The book tries to give a readable overview of some of the research that has helped to advance these themes in recent years, and thus has shaped biological physics. The book focuses on the application to microbes—mostly bacteria—because the microbial world is such a wonderful laboratory for studying how living cells solve physical problems. Bacteria are easy to grow and to manipulate genetically. Many of their regulatory behaviors, physical structures and functions, and other properties are, although complex, still much simpler than in higher organisms. They are a starting point for testing quantitative models. At the same time, the microbial world is so diverse and rich in special cases that it almost seems that every microbial behavior that is not forbidden does occur in at least some instances. For these reasons many of the biophysical topics that are presented in this book were first developed and demonstrated in bacteria, either in familiar model organisms such as *Escherichia coli* or *Bacillus subtilis*, or in other species less well known.

In earlier epochs (a couple of decades ago) biological physicists certainly had the will to construct quantitative physical models for living systems. The enduring fascination with Kleiber's law, discussed briefly in chapter 1, attests to that.

Unfortunately, upon approaching the molecular level, experimental detail often was lacking and the trail went cold. What has changed is that tools are now available for constructing and testing molecular-level models for biological function. There has also been an influx of physically and mathematically enthused scientists applying these tools. Experimentalists can engineer new genes and circuits into living cells, isolate individual cells and count protein or other molecules within those cells, test stimulus and response, and compare the results to quantitative models and simulation. This book gives an idea of how physically motivated experiments and models are leading to a highly quantitative understanding of the living cell. Much of this understanding can be applied not only to bacteria but also more widely in eukaryotic biology and in particular to biomedical problems such as cancer.

The book arises from an introductory course in biological physics that I have taught for advanced undergraduates in the physical sciences at the University of Florida. The course has evolved over the years, gradually increasing its focus on the above themes in physical biology. Students want exposure to real scientific literature in this area, presented at a level that is accessible to the interested but non-expert physical scientist. Therefore this book attempts to walk the reader through some of the highlights of the literature. It is aimed at advanced undergraduates, beginning graduate students, and others with varying levels of expertise in biology, chemistry, and even physics. Aside from a very basic undergraduate science background, its only significant prerequisite is a willingness to use mathematics, in particular calculus and probability.

The first chapter begins with a very general overview of the microbial world and some terminology, followed by a consideration of size and energy constraints that impact bacteria. The second chapter introduces some basics of bacterial growth that have motivated recent investigations of resource allocation and heterogeneity in cell division. Subsequent chapters discuss gene regulatory networks and feedback, the role of noise or stochasticity in gene expression, and their application to phenotypic switching. Bacterial communication, in particular pheromone signaling and the very new topic of electrical signaling, is then discussed. The final chapter is a case study that involves nearly all of these topics: the sporulation/competence decision circuit in *B. subtilis*. This final example shows how a cell can integrate feedback, noise, heterogeneity, and chemical communication in order to control a phenotypic switch.

Again, the book does not attempt to be comprehensive. Many interesting bacterial behaviors such as motility (including chemotaxis), social behavior (competition and cooperation), and evolution are only briefly mentioned. Furthermore, needless to say, this is not a biology textbook. Although I have tried not to get any biology wrong, I have not tried to explain it authoritatively either. That is the job of a microbiology textbook. Nevertheless, where biological jargon arises that may leave the physical scientist reader at a disadvantage, I have tried to explain it in a footnote.

A few exercises are sprinkled throughout. Most are designed to fill in mathematical or physical details that would bog down the discussion. They vary in difficulty.

Finally I thank the students who have sat in my classes for their interest and enthusiasm. I apologize in advance for any mistakes or serious omissions and accept responsibility for them.

Acknowledgements

I did not always study microbes or biological physics. I started out in a very different area of physics—experimental condensed matter physics—and only more recently began developing an interest in the physical biology of microbes. I am grateful to the colleagues, collaborators, and graduate and undergraduate students who helped me to do this. I especially thank Jonathan Young, Minjun Son, Robert Burne, Sang-Joon Ahn, Max Teplitski, and Eric Stabb for all the different kinds of help, support, advice and patience they were able to offer as my lab and I transitioned into this wonderful area.

Author biography

Stephen J Hagen

Steve Hagen is Professor of Physics at the University of Florida. He began his scientific career in experimental condensed matter physics, studying high temperature superconductivity. He then switched to biological physics, studying first protein folding dynamics and then bacterial communication. Most recently his research work has focused on unraveling noisy regulatory pathways in quorum sensing bacteria. He has published numerous scientific papers in all of these areas, as well as book reviews of physics textbooks for professional journals, and he recently edited a book on the physics of bacterial communication. His interests also include physics education and public policy. Prior to working at the University of Florida he worked at the National Institutes of Health in Bethesda Maryland and he was also a congressional science fellow with the American Institute of Physics.

IOP Concise Physics

The Physical Microbe
An introduction to noise, control, and communication in the prokaryotic cell
Stephen J Hagen

Chapter 1

Introduction

1.1 Diversity

A microbe is a microscopic organism, too small to be seen with the unaided eye. This definition is very broad. It encompasses a huge diversity of species that differ vastly in their evolution, size, shape, structure, physiology, habitat, and more. The differences between these organisms are so wide ranging that the term microbe is not very informative. We should distinguish between several kinds of life that can all be called microbes. At the highest level is the distinction between cell-based life such as bacteria and non-cellular agents such as viruses, phages, viroids, and prions. At the next level, cell-based life can be categorized in various ways. But, while it is tempting to classify cells according to their size, morphology, and other visible properties, the definitive classification of cellular life comes from phylogenetics, or evolutionary history, especially as revealed by comparison of DNA sequences. In particular, the DNA sequence that encodes a piece of ribosomal RNA known as 16S provides the basis for the widely accepted grouping of all cell-based life into three major domains. Using 16S RNA sequencing, all living cells can be placed within one of three domains: Bacteria, Archaea, or Eukarya. All three domains contain microbes. Eukarya includes the kingdoms of the protists (including protozoan microbes such as *Amoeba*) and fungi (including the common, single-celled baker's yeast *Saccharomyces cerevisiae*) as well as larger plants and animals. Bacteria includes many thousands of single-celled species, from the famous (or infamous) such as *Escherichia coli*[1] to the utterly obscure. Archaea, like Bacteria, are single-celled. They physically resemble bacteria but differ significantly in some key biochemistry and metabolic pathways. Archaea includes a number of extreme-loving

---

[1] A particular type of organism is identified by its *Genus* (capitalized italic) and *species* (lowercase italic). *Escherichia* is a genus or group of closely related bacteria and *coli* refers to one species within that group. After the organism has been mentioned once, its name is normally abridged as the first initial of the genus plus the name of the species, as in *E. coli*.

doi:10.1088/978-1-6817-4529-9ch1   1-1

(extremophile) species with high tolerance for heat (thermophiles), salt (halophiles), or radiation (radiophiles).

Non-cellular life forms do not fall within any of the three domains as they lack organelles and ribosomal RNA. Viruses for example are microscopic packages of nucleic acid that, while able to infect living cells, lack any independent ability to replicate. They depend on cellular life for this function.

Despite the all encompassing nature of the term microbe, this book will focus almost entirely on bacteria, with occasional mention of certain eukarya and archaea. The reason is that it aims to explore some of the physical problems that unicellular organisms have to solve, and bacteria alone comprise tens of thousands of examples of species that solve these problems in different ways. Among bacteria there is huge diversity, with different species inhabiting virtually every type of habitat and condition, and surviving and competing by almost every imaginable (or not) mechanism. In this book we will discuss, at least in passing, bacteria that are native to a wide range of habitats, from soils, marine sediments, and fish intestines, to insect digestive tracts, infected wounds, and the human oral cavity[2].

In addition to phylogenetic classifications, other types of distinctions are also useful in dealing with microbes. Cells can be classified structurally as either eukaryotic (having membrane-bound organelles such as the nucleus) or prokaryotic (lacking such organelles). Then archaea and bacteria are prokaryotes while eukarya are eukaryotic. Bacteria are often also classified by morphology, as either spherical (cocci), rod-like (bacilli), or spiral. An important practical distinction among bacteria is that between Gram-positive and Gram-negative species. The names refer to whether the bacterial cells can be dyed (and hence studied under a light microscope) using a crystal violet stain[3]. This would seem a completely arbitrary classification, except that it reveals whether the cells possess a thick layer of peptidoglycan (a polymer of sugar and amino acids) in their cell wall. Gram-positive bacteria (including *Streptococcus* and *Staphylococcus*) have a thick peptido-glycan layer, whereas Gram-negatives (such as *E. coli*) have a thinner peptidoglycan layer, between an inner and outer membrane. The crystal violet binds to the peptidoglycan, turning Gram-positive cells violet. Although the Gram staining is not as robust a classification as those based on ribosomal sequencing—some species do not fall cleanly under either the Gram-positive or Gram-negative classification—it is still an important empirical distinction. The different cell wall composition of Gram-positives and Gram-negatives has implications for the permeability of the cell wall, susceptibility to certain antibiotics, and other practical matters.

1.2 Size

Being single-celled, microbes are invariably 'small'. Many familiar species such as *E. coli* have a diameter ~1 μm and a mass ~1 pg. But the term microbe still permits a tremendous range of sizes, as indicated in figure 1.1. The smallest prokaryotes

---

[2] The mouth.
[3] The method was developed by Hans Christian Gram.

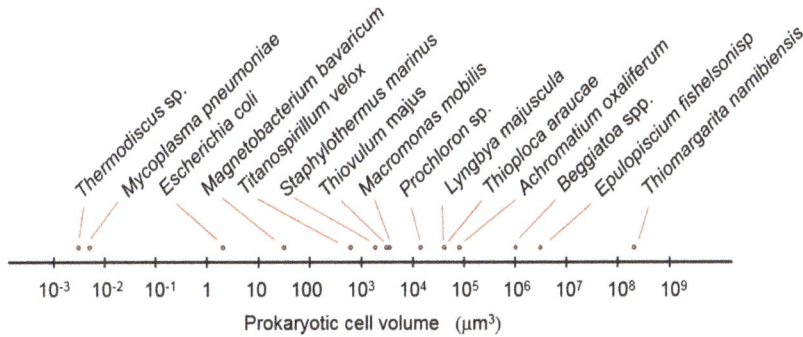

Figure 1.1. Prokaryotic organisms vary dramatically in size. Although some of the values are extremes measured for unusual individuals, other values are typical of the species. The very largest bacteria are either sulfur bacteria, which inhabit O_2-poor, sulfide-rich sediment environments, or cyanobacteria, the photosynthetic blue-green algae. The figure is adapted from table 1 of [1].

include both archaea and bacteria. The so-called nanobacteria include cells with diameters as small as 0.2–0.3 μm. *Thermodiscus* is a genus of disk-shaped archaea with diameters as small as 0.2 μm and a thickness 0.1–0.2 μm. At the other extreme from the nanobacteria are giant bacteria. The largest of these is *Titanospirillum namibiensis*, which was found on the sea floor off Namibia. These spherical sulfur bacteria attain a diameter of 750 μm and grow in chains, like a string of pearls. Consequently, over the full range from the smallest nanobacteria to the largest giant bacteria the cell volumes span from 0.01 μm³ to as much as 2×10^8 μm³, a full ten orders of magnitude. Since the giant bacteria must obtain nutrients more or less the same way that smaller bacteria do, by diffusion, their large size must affect the way that they gather nutrients to support their metabolism. We will return to this issue.

Bacteria are also small in terms of genome size. As table 1.1 shows, bacterial genomes are typically in the range of a few million DNA base pairs (1 Mb = one million base pairs[4]). This is a small number in comparison to 144 Mb (fruit fly, *Drosophila melanogaster*) or 3.2 Gb (human, *Homo sapiens*)[5]. Because each base pair (bp) contributes about 0.34 nm to the length of the DNA double helix, the chromosome of the bacterium is a DNA molecule with an outstretched length near 1 mm, packed into a micron-sized cell.

The rule of thumb for bacterial genomes appears to be that they contain about one protein-coding gene for each kilobase (kb) of DNA, with the details depending on the particular bacterium and strain: *E. coli* strain O157:H7 has about 5.5 Mb with 5400 genes, while strain K-12 has 4.6 Mb and about 4500 genes. *Mycoplasma genitalium* survives with a tiny chromosome of 0.58 Mb, with an impressively small set of about 570 genes. Some bacteria have much smaller genomes. This naturally raises the question of what is the smallest possible genome, or the smallest number of

---

[4] Recall that the DNA double helix consists of two long strands of nucleotide bases. The four types of nucleotide bases are denoted A, T, C, G. Bases along one strand are paired with complementary bases on the opposite strand to give the double helix structure.

[5] However, the higher organism genomes do not contain proportionally more protein-coding genes.

Table 1.1. Some cell size and metabolic data, including mass-specific endogenous metabolic rates q (W kg^{-1}), for various prokaryotes. The mass and q data are from [7]. Genome size data are from NCBI. The mass shown is the wet cell mass, which is related to the cell volume by a density that is very nearly 1 μm^3 pg^{-1}. As aerobic metabolism generates about 20 J of energy for each 1 ml O$_2$ consumed, these metabolic rates are estimated from the rate of oxygen consumption of a live culture, normalized to its dry mass, and multiplied by the ratio of dry to wet cellular mass, about 0.3.

Organism	Mass (pg)	Genome (Mb)	Endogenous q (W kg^{-1})
Francisella tularensis	0.01	1.9	3.7
Mycobacterium tuberculosis	0.2	4.4	7
Staphylococcus aureus	0.27	2.7	2.7
Streptococcus pneumoniae	0.4	2.0	1.7
Vibrio fischeri	0.6	4.3	22
Escherichia coli	0.7	5.4	17
Bacillus subtilis	1.4	4.2	6.5
Sphaerotilus natans	6.5	4.6	45

genes that an organism can have? *M. genitalium* has the smallest genome known among organisms that can grow independently of other species, in culture medium. But it does not possess the smallest genome by any means. A bacterium that lives in symbiosis with other species can manage with a smaller subset of genes, mostly those required for the core processes of DNA replication, transcription, and translation.

For example, many of the so-called obligate symbionts of insects—bacteria that live exclusively in symbiosis with insect hosts—have extremely small genomes. The smallest known at present is *Candidatus Tremblaya princeps* which has a genome of 0.139 Mb, or only 121 recognizable genes. Such bacteria may inhabit the digestive systems of insects that provide a steady stream of food such as blood or plant sap. The bacteria retain genes that support the symbiosis, for example synthesizing amino acids that their host animal cannot obtain from its diet. But they may lack not only the genes needed to feed themselves, but also many other important genes. These bacteria do benefit from their symbiotic lifestyle, but they consist of small, isolated populations that reproduce asexually within their host animals. As a result they tend to accumulate genetic defects through irreversible and inevitable mutations and gene losses. These defects lead to deficiencies in basic functions such as cell envelope biosynthesis, DNA repair and replication, and even in the proper folding of such proteins as they can still manufacture [2].

1.3 Energy

Another important microbial classification is based on metabolism. Consistent with their different lifestyles and habitats, microbes show great diversity in the ways that they satisfy three metabolic needs. Living cells require (1) carbon for synthesis of protein and other biomolecules, (2) electrons or other reducing species to carry out respiration and biosynthetic reactions, and (3) energy. They can be classified by the way they meet these needs:

1. Some microbes obtain their carbon directly from CO_2. These are known as autotrophs. Those that obtain carbon from organic matter are known as heterotrophs.

2. Many familiar microbes obtain electrons or reducing equivalents by oxidizing organic material. These are known as organotrophs. Many other species, all of which are either bacteria or archaea, obtain electrons from inorganic substances such as hydrogen (H_2), sulfur, or sulfide (S^{2-}). These are known as lithotrophs. Lithotrophs may use O_2 as their electron acceptor or they may use another chemical substance such as sulfate (SO_4^{2-}).

3. Some microbes, known as phototrophs, obtain energy through photosynthesis. Those that obtain energy by oxidizing chemical species in their environments are chemotrophs.

The nomenclature above gives rise to additional subclassifications such as photo-autotrophs, which acquire their energy and carbon from sunlight and CO_2, and chemolithoheterotrophs, which gather energy by oxidizing inorganic compounds but obtain carbon from organic matter. Examples will turn up shortly.

Although a small organism can only consume a small amount of energy, bacterial populations can be large and have large collective impact. Heterotrophic bacteria may account for 50%–90% or more of the total respiration (oxygen-consuming metabolism) that occurs in some ecosystems, such as in the oceans [3]. In these environments bacteria are consuming the overwhelming share of the 'primary productivity', the energy made available by photosynthesis.

How much energy does an individual bacterium consume? The rate of energy consumption by any living cell depends on the cell's function and its stage of growth. Cells in rapid growth require more energy. To make useful comparisons between organisms one compares their rate of energy consumption in a resting state. For mammals and birds this is known as the basal metabolic rate. For single-celled organisms it is the endogenous metabolic rate, which is the rate of energy consumption in the non-growing state typical of low-nutrient conditions. The organism's metabolic rate is denoted Q (W, watts). Since even unicellular organisms vary widely in size (table 1.1), it is customary to normalize Q to the organism's mass M, giving a mass-specific metabolic rate $q = Q/M$ (W kg^{-1}) that can be compared for species on different size scales.

Within many groups of related organisms, such as ectothermic (cold-blooded) vertebrates, Q is found to scale as a power law in body mass $Q \propto M^{3/4}$ and so $q \propto M^{-1/4}$, a famous result known as Kleiber's law [4, 5]. Although the 1/4 exponent in Kleiber's law seems small, it nevertheless predicts that the specific metabolic rates of unicellular organisms should be vastly greater than those of larger animals, as unicellular masses are tiny. For example, based on the specific metabolic rate of an Asian elephant [6] $M \simeq 4 \times 10^6$ g, $q = 0.6$ W kg^{-1}, Kleiber's law predicts that a microbe such as *E. coli* with $M \sim 10^{-12}$ g will have $q \sim 30\,000$ W kg^{-1}.

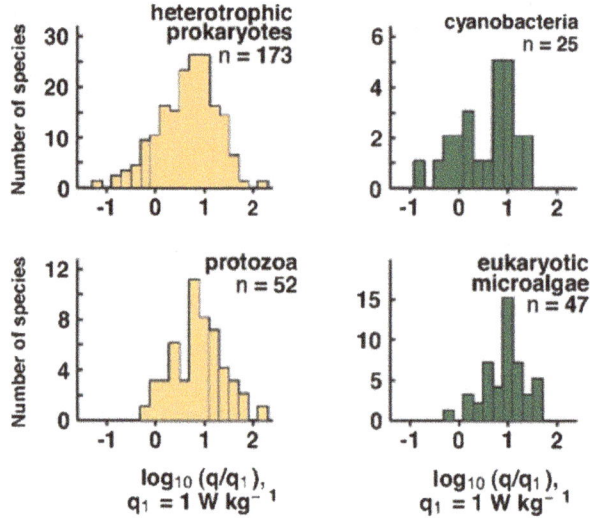

Figure 1.2. Histograms of mass-specific, endogenous metabolic rates q for heterotrophic (left) and photo-autotrophic (right) single-celled microbes. Although there is considerable variability within each group, q generally falls within the range 0.1–3 W kg^{-1}. Reproduced with permission from [7]. Copyright (2008) National Academy of Sciences, USA.

That prediction is, however, incorrect. The values of q for single-celled organisms do vary over a very wide range, as shown by figure 1.2. But, setting aside the extreme values, the typical unicellular q are not so different from those of much larger animals. In a survey [7] of 3006 aerobic species that included 173 species of prokaryotes as well as many algae, plants, and invertebrate and vertebrate animals, the typical values of q within each subgroup of organisms invariably fell within a narrow range of about 30-fold, about 0.3 W kg^{-1} to 9 W kg^{-1}. That is, the slowest-metabolizing bacteria are not so different from animals such as crustaceans and ectothermic (cold-blooded) fish, which have $q \simeq 0.1$–0.3 W kg^{-1}. The fastest-metabolizing prokaryotes are comparable to endothermic vertebrates (birds and mammals), which average around 1–3 W kg^{-1}. Table 1.1 lists some specific examples among microbes. Again, these figures describe endogenous metabolic rates. In actively dividing bacteria the highest values of q are much higher, of order 10^3 W kg^{-1}. These values are similar to those of flying or hovering insects and birds, giving an idea of the energy demand of exponential growth[6].

Figure 1.3 shows one unexpected feature of the data. The endogenous q shows no statistically significant scaling with respect to M among the unicellular organisms. Although the $q \sim M^{-1/4}$ law holds within groups of multi-cellular organisms it does not seem to hold among microbes.

---

[6] Compare this power consumption to that of an incandescent light bulb. A 60 W light bulb of mass 40 g has a specific metabolic rate $q \simeq 1500$ W kg^{-1}.

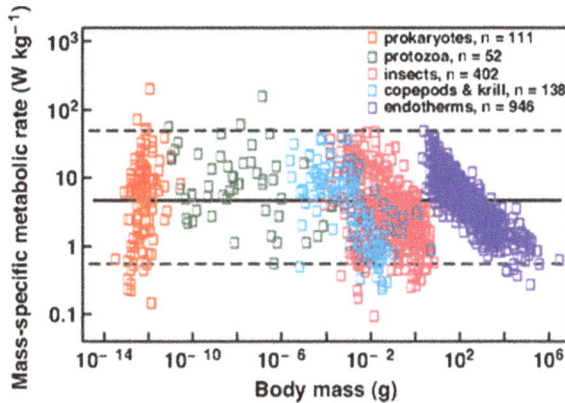

Figure 1.3. Mass-specific basal metabolic rates and body masses for a range of organisms, from microbes (prokaryotes and protozoans) to large warm-blooded animals (endotherms). Although some trends are visible within some groups, most species nevertheless lie within a range of about 0.3–30 W kg^{-1}. Reproduced with permission from [7]. Copyright (2008) National Academy of Sciences, USA.

1.4 Food

Not having respiratory systems or digestive tracts, microbes must transport O_2 and nutrient from the environment across the plasma membrane and into the cell. O_2 and other small, electrically neutral and non-polar molecules pass readily through the membrane by simple diffusion. Ions and larger or polar molecules such as peptides or sugars cannot easily pass through the hydrophobic interior of the membrane and so must be assisted by dedicated and often molecule-specific mechanisms. This may be accomplished by facilitated diffusion, in which membrane proteins such as channels permit an extracellular nutrient molecule to transit the membrane. This form of transport is ultimately driven by the concentration gradient between the cell and its surroundings and does not require an input of energy.

Alternatively, transport may occur through an active (energy-consuming) mechanism. ATP-binding cassette transporters (ABC transporters) are an important example found widely across the three domains of life. In an ABC transporter a protein, perhaps residing on one surface of the plasma membrane, binds or captures a particular target molecule and then brings it to a transporter protein that is embedded in the membrane. The transporter pushes the target across the membrane with energetic assistance from an ATP-consuming reaction that occurs on the intracellular side of the transporter. Because it consumes energy, the ABC transporter can drive a flux of target molecules up a concentration gradient, causing a nutrient to accumulate in the cell even though it is scarce in the environment. By contrast passive diffusion only allows flux down the concentration gradient.

Group translocation is an active transport mechanism in which the molecule to be transported is chemically modified for the transport process. The sugar phosphotransferase systems used by many bacteria are an important example: sugars like glucose and fructose are phosphorylated through a process involving a series of enzymes prior to import into the cell.

The narrow range of metabolic rates shown in table 1.1 gives an idea of the nutrient or O_2 flux that a microbe of given size will require. Aerobic metabolism requires approximately 1 ml O_2 for every 20 J of energy produced by respiration. Therefore an actively dividing *E. coli* may consume 2.4×10^{-7} μmol O_2 per day, while in its slow-growth state (or stationary phase, to be discussed later) it may consume 3.3×10^{-10} μmol O_2 per day [8]. This is mostly delivered by diffusion. If O_2 or an essential nutrient is present in the environment at concentration c_0, and the organism captures the molecules that diffusion brings to its exterior surface, then diffusion will deliver to the cell a steady flux of nutrient at a rate that depends on the size of the cell, the average concentration c_0 of nutrient in the environment, and the diffusion coefficient D of the nutrient. In the case where a spherical cell (radius a) absorbs every nutrient molecule that makes contact with its surface, the total flux (particles/time) of that nutrient into the cell is given by the diffusional rate $k = 4\pi Dac_0$.

At the micron scale of a bacterial cell, diffusion allows for rapid and efficient mixing of small molecules. However, it is highly inefficient on longer length scales. Typical small molecules have diffusion coefficients $D = 10^{-5}–10^{-6}$ cm^2 s^{-1}. Therefore an oxygen molecule diffuses about 1 mm in 1 h, but only about 2 cm in 1 day. Diffusion alone is certainly too slow to deliver a useful flux of food or oxygen over distances of centimeters or more. If food is distant, it is more efficient for the cell to swim toward it. Many bacteria are equipped with flagella as well as complex sensory mechanisms that allow them to swim up concentration gradients toward nutrients.

Down at the micron scale, diffusion of oxygen or nutrient into the cell creates a nutrient-depleted zone that surrounds and follows the cell. The size of this zone is determined by the size of the cell and the efficiency of uptake at the cell boundary. The zone width is similar to the size of the cell, not more than a few micrometers. If the food is a distance l away, then the time required for it to diffuse toward the cell is roughly $t_{\text{diffuse}} \simeq l^2/D$, while the time required for the cell to reach the food by swimming at speed v is $t_{\text{swim}} \simeq l/v$. Swimming is then advantageous only if $t_{\text{diffuse}}/t_{\text{swim}} = vl/D > 1$, or $v > D/l$. That is, swimming or stirring is helpful only if Pe > 1, where Pe $= vl/D$ is the dimensionless ratio known as the Péclet number[7]. If the nutrient is $l = 1–10$ μm away and diffuses at $D = 10^{-5}$ cm^2 s^{-1}, then the organism must move with $v > 0.1–1$ mm s^{-1} to outrun diffusion. Few species can swim this fast. Bacterial swimmers are typically in the league of *E. coli*, which swims at about 30 μm s^{-1}. There are exceptions, however. *Thiovulum majus* is an unusual, large (5–25 μm diameter) sulfur-consuming bacterium that swims at speeds as great as 600 μm s^{-1} through oxic/anoxic interface zones where it can oxidize hydrogen sulfide [9]. For most bacteria, however, if food is only a few micrometers away, it is easier simply to wait for it to arrive by diffusion than to try to outrun diffusion and actively swim to the food, or to attempt to stir the local environment.

---

[7] The Péclet number 'Pe' describes the ratio of advective (stirring) to diffusive transport in a fluid.

1.5 Diffusion versus size

The nutrient flux $k = 4\pi Dac_0$ delivered by diffusion alone increases in proportion to a, the radius of the cell. However, the organism's energy needs scale in proportion to its volume and its specific metabolic rate, $k \propto qa^3$. Therefore the energy need increases as a^3 while the nutrient flux increases only as a. As a cell becomes larger its energy needs grow more quickly than the rate at which nutrient arrives. Since there seems to exist a lower limit to the metabolic rate, this argument suggests that there exists a natural upper limit on the possible size of the bacterium. The bacterial size at which the nutrient flux exactly matches the metabolic rate will depend on the specific metabolic rate, $a \propto q^{-1/2}$. For a typical cell with $a \simeq 1$ μm and a value $q \simeq 1$ W kg^{-1} at the upper range in table 1.1, the range of plausible q suggests it should be possible for the cell to have about a 30-fold smaller value of q. It should then be possible for the cell to be about $\sqrt{30}$ times larger, or about $a \simeq 5$ μm.

An additional and related constraint arises from the fact that bacteria have no cytoskeleton for active transport of molecules inside the cell. They rely on diffusion to provide chemical mixing in the cytoplasm. One can estimate the time required for two molecules, each of size R, to find each other inside a cell of size a. If both molecules have a diffusion coefficient D, their rate of interaction is roughly the intracellular concentration of the first molecule ($c_o \simeq 1/a^3$) multiplied by the diffusion-controlled rate of its encounter with the second molecule, $4\pi DR$. The total rate of interaction is then roughly DR/a^3. The time scale for an interaction is then $\sim a^3/RD$. If the cell size is about $a \sim 1$ μm and the molecular size is about $R \sim 10^{-6}$ cm, this interaction time is a manageable 0.1 s; a small microbe is internally well-stirred.

However, if the cell is much larger, such as $a \sim 100$ μm, the interaction time increases by six orders of magnitude to almost 30 h. In such a large cell, chemical mixing will be very inefficient and chemical heterogeneities or micro-environments will develop. These are likely to interfere with the efficient regulation of all the important biochemical processes, protein synthesis and so forth. Consequently, unless the nutrient environment is rather unusual one anticipates that bacteria should not grow larger than the diffusion-limited size estimated above, about 5 μm, with a maximum volume of about $a^3 \sim 100$ μm^3.

This estimate turns out to be generally correct, as table 1.1 suggests. However, figure 1.1 shows that some bacterial species do exceed this size limit. The largest archaea seems to be *Staphylothermus marinus*, a sulfur-consuming thermophile that can reach 15 μm. The absolute largest known bacterium is the *T. namibiensis* mentioned earlier, which attains a diameter of 750 μm and is easily visible to the unaided eye.

How do these species evade the diffusion limit? When nutrient is scarce smaller cells should have an efficiency advantage. But the largest bacteria generally occupy unique nutrient environments and employ some physiological problem-solving that together bend the rules in favor of larger size [1]. Many of them are either sulfur bacteria, which obtain energy by oxidizing hydrogen sulfide, or else cyanobacteria, which obtain energy through photosynthesis. The largest bacterium, *T. namibiensis* is a chemolithoautotrophic sulfur bacterium. It uses O_2 or nitrate as electron acceptors to oxidize H_2S. It inhabits marine sediment zones where O_2-containing seawater comes into contact with H_2S-rich, but O_2-poor, sediment. Despite its large cell size, however, some 98% of its cellular volume is occupied by large vacuoles filled with nitrate. The active cytoplasm comprises only a very small volume around the perimeter of the cell. Its nitrate reservoir allows the cell to continue oxidizing H_2S during the long intervals when the bacterium, lying immotile in the undisturbed sediment, has plenty of sulfide nearby but no access to O_2.

Beggiatoa spp.[8] are also large sulfur bacteria, with cell sizes up to 200 μm, that inhabit marine sediment surfaces where they live by oxidizing H_2S with O_2.[9] As they require access to both, they inhabit the oxygen/sulfur interface, where opposing gradients of oxygen (from the water) and sulfide (from the sediment) meet. *Beggiatoa* form highly active layers on the sediment surface, consuming O_2 at a diffusion-limited rate from the bulk water above and generating a strong O_2 concentration gradient. *Beggiatoa* assemble into long filaments of hundreds or thousands of cells, and these filaments can actively swim to find the ideal chemical gradient conditions. They are able to transport nitrate down from the water interface to oxidize hydrogen sulfides further down in the sediment. Like *T. namibiensis* they possess large vacuoles that can store nitrate for oxidation of sulfide during intervals when oxygen is unavailable.

Finally, *Epulopiscium fishelsoni* is another very large heterotrophic bacterium. Reaching a diameter up to 80 μm and a length of 600 μm, it lives in the gut of a fish found in the Red Sea. It is plausible that its large size is enabled by an unusually high concentration of nutrients in the fish gut. In addition, *E. fishelsoni* is rod-shaped. Presumably the rod shape is an optimal strategy for increasing surface area (relative to volume) and maximizing food uptake by diffusion. But as diffusion is insufficient to distribute its gene products efficiently throughout the cytoplasm, *E. fishelsoni* seems to possess a network of vesicles, tubules and other structures that may serve a transport function. Remarkably, each cell also carries many hundreds of copies of the genome. It defeats the chemical mixing problem by spatially distributing its gene regulation and protein synthesis functions.

Exercise (Diffusion of a nutrient) *Suppose a nutrient is present in a solution at a concentration c_0 (particles per unit volume) and it diffuses toward a spherical cell (radius a) with a diffusion coefficient D.*

- *The concentration at a distance* r *from the center of the cell must be a function $c(r, t)$ where $c(r, t) \to c_0$ as $r \to \infty$. $c(r, t)$ obeys the diffusion equation,*

$$D\frac{1}{r^2}\frac{\partial}{\partial r}\left(r^2\frac{\partial c}{\partial r}\right) = \frac{\partial c}{\partial t},\tag{1.1}$$

---

[8] Terminology: 'sp' (singular) or 'spp' (plural) means 'species'. *Beggiatoa* sp and *Beggiatoa* spp refers to one species and several species, respectively, of the genus *Beggiatoa*.
[9] *Beggiatoa* are chemolithoheterotrophs, to be precise.

but in the steady state, where $c(r, t)$ is unchanging over time, $\partial c/\partial t = 0$, so that c is a function of r only, $c(r)$. Suppose that every nutrient molecule that reaches the cell surface is absorbed immediately, causing $c(a) = 0$. Find $c(r)$ for $r > a$.

- *The diffusional flux of particles passing through unit cross-sectional area per unit time is j (particles/area/time). The flux j in the radial direction (toward the cell) can be found from $c(r)$ using Fick's law of diffusion,*

$$j(r) = -D\frac{\partial c}{\partial r}. \tag{1.2}$$

Note that the radial flux j is a function of r. Find the steady state flux of particles to the cell as a function of D, a, c_0. (That is, verify that the flux is as stated above, $4\pi Dac_0$.)

- *Now suppose that the surface of the cell absorbs the nutrient at a first order rate k, meaning that the number of particles absorbed per unit time is $kc(a)$ where $c(a)$ is the nutrient concentration at the surface of the cell ($r = a$). Find $c(r)$ in this case.*
- *For the latter case, find the steady state flux of particles to the cell as a function of D, a, k, c_0.*

References

[1] Schulz H N and Jorgensen B Barker 2001 Big bacteria *Ann. Rev. Microbiol.* **55** 105–37

[2] McCutcheon J P and Moran N A 2012 Extreme genome reduction in symbiotic bacteria *Nat. Rev. Microbiol.* **10** 13–26

[3] Rivkin R B and Legendre L 2001 Biogenic carbon cycling in the upper ocean: effects of microbial respiration *Science* **291** 2398

[4] Kleiber M 1932 Body size and metabolism *Hilgardia* **6** 315–53

[5] Gillooly J F, Brown J H, West G B, Savage Van M and Charnov E L 2001 Effects of size and temperature on metabolic rate *Science* **293** 2248–51

[6] Makarieva A M, Gorshkov V G and Li B-L 2005 Energetics of the smallest: do bacteria breathe at the same rate as whales? *Proc. Biol. Sci.* **272** 2219

[7] Makarieva A M, Gorshkov V G, Li B-L, Chown Steven L, Reich Peter B and Gavrilov Valery M 2008 Mean mass-specific metabolic rates are strikingly similar across life's major domains evidence for life's metabolic optimum *Proc. Natl Acad. Sci.* **105** 16994–9

[8] Riedel T E, Berelson W M, Nealson K H and Finkel S E 2013 Oxygen consumption rates of bacteria under nutrient-limited conditions *Appl. Env. Microbiol.* **79** 4921–31

[9] Fenchel T 1994 Motility and chemosensory behaviour of the sulphur bacterium *Thiovulum majus Microbiology* **140** 3109–16

The Physical Microbe
An introduction to noise, control, and communication in the prokaryotic cell
Stephen J Hagen

Chapter 2

Growth

2.1 Exponential growth

If bacteria are known for one thing, other than their small size, it is their rampant growth. Bacteria increase in number by repeated cell division, with each cell growing in size over a period of time and then dividing into a pair of daughter cells. Although one usually envisions the parent cell dividing into two identical daughters, not all bacteria divide symmetrically. The parent cell of *Caulobacter crescentus* divides into two morphologically different daughters. One is an adherent, sessile cell that continues to further divide. The other daughter is motile and swims away. *Bacillus subtilis* can divide in a process of sporulation whose regulation is discussed at length in a later chapter. Sporulation combines formation of a spore daughter with lysis[1] of the parent. Species of the genus *Epulopiscium*, one of the giant bacteria discussed in the previous section, reproduce exclusively in a sporulation process, with multiple daughters forming within the parent cell. The parent eventually ruptures and releases the daughters.

Consider, however, the more familiar division behavior of *Escherichia coli* where the parent divides into two near-identical daughters. If a culture starts with n_0 cells at time $t = 0$ and the cells divide repeatedly at equal time intervals of τ_d, then the population $n(t)$ increases as

$$n(t) = n_0 2^{t/\tau_d}. \tag{2.1}$$

Under the most favorable conditions some strains of *E. coli* can divide as frequently as $\tau_d = $ 15–20 min, although slower growth with $\tau_d \sim 1$ h is typical for many species growing in laboratory conditions. Equation (2.1) is equivalent to

$$n(t) = n_0 \exp(\lambda t), \tag{2.2}$$

---

[1] A cell undergoing lysis loses membrane integrity and releases its contents to the environment.

doi:10.1088/978-1-6817-4529-9ch2

where $\lambda = \log 2/\tau_d$ is known as the growth rate[2]. Note that (2.2) implies that the population increases by a fraction $dn/n = \lambda dt$ in every short time interval dt. That is, equation (2.2) solves the differential equation

$$dn = \lambda n \ dt \rightarrow \frac{dn}{dt} = \lambda n. \tag{2.3}$$

The growth rate λ depends on nutrient, temperature, and other environmental conditions as well as the species and strain being studied. It can become limited by the accumulation of metabolic waste products as a population grows, or by changes in the ionic conditions such as pH. Laboratory studies are often designed so that λ is limited by the scarcity of a particular component or essential nutrient such as glucose. This makes the growth rate sensitive to the concentration of that nutrient and it allows investigation of the biochemical pathway for nutrient uptake and processing. In such cases λ typically increases in proportion to the concentration c of the limiting nutrient, and it saturates when c is large enough that it is no longer the limiting factor for growth. This saturation tends to follow the empirical law stated by Jacques Monod in 1949,

$$\lambda = \lambda_0 \frac{c}{c_1 + c}, \tag{2.4}$$

where c_1 and λ_0 are constants [1].

Growth rate also accelerates with temperature, following a mostly Arrhenius-like behavior,

$$\lambda = A \exp(-\Delta E/k_B T), \tag{2.5}$$

at least within about 10 °C–20 °C of the natural growth temperature. Here ΔE is an activation energy, perhaps $(7-12) \times 10^{-20}$ J $= 40$–70 kJ mol^{-1}. At lower temperatures, however, growth rates tend to fall short of what (2.5) predicts [2].

Surprisingly, the rate of division is not limited by the speed of DNA replication. Cell division can occur in a shorter time than it takes to replicate the chromosome. The organism accomplishes this by having several rounds of replication in progress at one time. That is, it may have multiple copies of its chromosome, all in different stages of completion, and therefore also multiple copies of any one gene.

Rapid, exponential growth can allow a few colonizing bacteria to invade and overrun a food source in just a few hours. It also generates significant adaptive power, as even a very small incidence of error in DNA duplication inevitably creates enormous numbers of mutants in a population that increases exponentially. For example the E. coli chromosome consists of about 5×10^6 or roughly 10^7 DNA base pairs (~10 Mbp), depending on the particular strain. During cell division these are copied with a low error probability of only 10^{-9} per base pair copied. If all errors are equally likely, then with each division an error will appear in the chromosome of the

---

[2] Although (2.2) describes exponential growth, microbiologists usually refer to this behavior as logarithmic growth. This is perhaps an archaism from the era when growth curves $n(t)$ were plotted by hand on logarithmic graph paper.

daughter cell with probability $p \sim 0.01$. But if all such mutants survive and reproduce in a cell culture that grows overnight, propagating their mutations to their daughter cells through 30 or so generations, then the resulting culture will contain 10^9 cells, each of which has a probability $\sim 1 - (1 - p)^{30} \simeq 0.26$ of carrying at least one mutation. Hence the number of mutants in the culture is predicted to be $\simeq 2.6 \times 10^8$. Surprisingly this number exceeds the number of distinct single base-pair mutations that are even possible in a 10 Mbp chromosome. This naive argument suggests that, following a night of uninhibited growth from a single cell, every possible single-base DNA mutation will probably arise at least once and be represented in the culture.

Every possible mutation does not appear, however, because different mutations bring small advantages or disadvantages in growth rate. Faster growing mutants will easily outcompete the others over 30 generations. In short, rapid growth favors genetic drift in a bacterial population. For this reason, laboratory-created mutations are not necessarily retained in a strain that is cultured repeatedly ('passaged') through many generations in the laboratory. Similarly, bacterial strains collected in natural habitats can begin to differ significantly from their wild-type cousins after they have been passaged many times in the laboratory. Repeated passage under stable, well-controlled laboratory conditions can lead to loss of genes and mechanisms that are of high value to the wild-type organism in its more variable natural environment.

2.2 Stationary phase

Exponential growth never continues indefinitely. If it did, a single bacterium of mass $\simeq 3$ pg could grow and divide unimpeded at intervals of $\tau_d \simeq 20$ min, creating after 24 h roughly 5×10^{21} progeny cells with a total mass 10^{10} g. Within the following day the total mass of its progeny would surpass the mass of the Earth, 6×10^{24} kg. This is clearly unrealistic. Therefore exponential growth must be one of several possible phases in the life of a bacterial culture. It ends when space, nutrients, O_2, and other resources are consumed and waste products accumulate. As a bacterial culture begins to exhaust its medium, λ declines and growth eventually ceases. A culture that has passed through its active-growth phase is said to have entered the stationary phase. Cells in stationary phase are not dead and can normally be revived by transfer into fresh growth medium. However, under a microscope the stationary cells are seen to be typically smaller and more variable in appearance than the plump, uniformly sized cells of exponential phase.

The transition from exponential to stationary phase is reasonably well described by the so-called logistic growth law [3], a plausible and mathematically tractable modification of (2.3) that dates to 1838. The logistic law cuts off the exponential growth as the population $n(t)$ reaches a carrying capacity denoted K:

$$\frac{dn}{dt} = \lambda n(1 - n/K). \tag{2.6}$$

One may integrate this equation and find the expression for $n(t)$, given a starting population n_0. It does not give a perfect fit to real bacterial $n(t)$ data, but it does share their generally sigmoidal shape and, in particular, their exponential behavior at early t. Other mathematical models exist that can sometimes provide a more precise fit to real data [4].

Exercise (Population versus time)   *Integrate (2.6) to find n(t). Assume that the initial (t = 0) value is n_0*

2.3 Lag phase and decline

There is more to a bacterial growth curve than just the exponential and stationary phases, however. When stationary phase cells are transferred into fresh growth medium, they do not immediately begin to grow at their maximum λ. More usually they require a period of time to ramp up to exponential growth, in part because they lack the cellular machinery to grow rapidly right away. Hence one usually observes a delay in growth when a stationary culture is transferred into fresh medium. This slow-growth state is known as lag phase.

In addition, a stationary culture that spends sufficient time in depleted medium will begin to die, and its population density will truly decline as cells lyse. Consequently the typical bacterial growth curve (population versus time), if a dormant culture is inoculated into a fresh medium, may consist of multiple phases that include a lag phase of a few hours, an acceleration of growth that is followed by exponential phase, a slowing into stationary phase and (ultimately) a phase of decline or death. Not all of these phases can be observed in every experiment, and in some cases additional phases may be visible[3]. Although it is tempting to think of exponential phase as the only interesting phase, in fact the other phases have been the subject of considerable study and show evidence of interesting regulatory mechanisms at work. We will discuss the lag phase further in the context of bacterial persistence.

2.4 Balanced growth

If the environment is both favorable and stable, a culture can grow at constant λ indefinitely, with its cells in a sort of stationary state of perpetual growth and division. This condition is known as balanced growth, a steady state neatly defined by Campbell (1957) as follows: 'Growth is balanced over a time interval if, during the interval, every "extensive" property of the growing system increases by the same factor' [5]. Balanced growth is intriguing as it requires every cell to duplicate itself in every detail, including its reproduction machinery, accumulating neither a surplus nor a deficit of any constituent, at regular intervals of τ_d, for an indefinite period. It is difficult to imagine a human-made machine that could churn out perfect,

---

[3] Diauxie, for example, is a phenomenon in which two exponential phases occur in succession as the culture depletes one preferentially consumed nutrient from its medium and then switches over to consuming another one.

functioning replicas of itself, at regular intervals, without accruing either a shortage or surplus of any part. Bacteria however can do this.

Balanced growth has intrigued researchers because a small number of parameters, connected by simple relationships, describe many properties of the growing cells. Because the cells are dividing at regular intervals, every cell in the population is nearly the same age, has a similar past history, and consequently shares similar size and chemical composition. Key parameters of the dividing cells include λ and τ_d, as well as the size (length, volume, or area) of the newborn and recently divided cells.

Many relationships between growth, cell mass, and composition during balanced growth were collected by researchers such as Schaechter and co-workers [6, 7] using bulk samples. The classic method for studying bacterial growth was to measure either the 'bacterial density' (the dry weight of cells per unit volume of a culture) or the cell concentration (the number of bacterial cells per unit volume), typically by a measurement of the culture's optical density[4]. These classic experiments showed that, during balanced growth, the growth rate λ largely controls the size and composition of bacterial cells. For example, the mass of the average cell increases exponentially with λ while the ratio of RNA to protein in the cell increases in proportion to λ. The RNA content is significant because most of the cell's RNA is tied up, along with ribosomal protein, in the structure of the ribosomes[5]. A single *E. coli* cell contains tens of thousands of ribosomes, a number that increases with growth rate. Consequently the RNA content indicates the cell's investment in machinery for protein synthesis. Growth rate links the size of the cell with its molecular composition and its allocation of resources.

An intuitively appealing interpretation for such dependencies is that balanced growth results from an autocatalytic cycle, where the cell contains several constituents that manufacture each other. For example, one can imagine a two-component cycle consisting of ribosomes and protein, where ribosomes make other protein at a rate k_t (mass of protein produced per mass of ribosome, per unit time), while the other protein manufactures ribosomes at a rate k_r. This gives

$$\frac{dP}{dt} = k_t R, \tag{2.7}$$

$$\frac{dR}{dt} = k_r P, \tag{2.8}$$

where R and P are the mass of ribosomes and protein, respectively, in the culture. The above equations predict that

---

[4] Optical density is the measure of the amount of incoming light that is absorbed or scattered by a sample. It offers an easy and fairly accurate way of tracking cell concentration in microbial cultures.

[5] Ribosomes are molecular machines that read the genetic information that is contained in messenger RNA transcripts and use that information to assemble polymers of amino acids. That is, they translate DNA sequence information into physical protein. The ribosome is a structure made from RNA and protein together.

$$\frac{d^2 R}{dt^2} = \lambda^2 R, \tag{2.9}$$

$$\frac{d^2 P}{dt^2} = \lambda^2 P, \tag{2.10}$$

where $\lambda = \sqrt{k_t k_r}$. These equations are solved by exponential growth at rate λ,

$$P = P_0 \exp(\lambda t), \tag{2.11}$$

$$R = R_0 \exp(\lambda t). \tag{2.12}$$

Because R and P grow at the same exponential rate, the model describes balanced growth[6].

Combining the derivative of (2.12)

$$\frac{dR}{dt} = \lambda R \tag{2.13}$$

with (2.8) gives

$$\frac{R}{P} = \frac{k_r}{\lambda} = \frac{\lambda}{k_t}. \tag{2.14}$$

That is, the ratio of RNA to protein increases in proportion to λ, at least if k_t is constant. The autocatalytic model then reproduces the observations of Schaechter *et al* [6, 7] as long as the cell responds to changes in nutrient conditions by altering the rate of ribosome manufacture k_r, while keeping the translation rate k_t fixed. This simple model suggests that the protein output per ribosome does not change as growth conditions change. Rather the cell modulates λ by adjusting the number of ribosomes so as to keep each ribosome working at full capacity. When more nutrient is available, the cell ramps up growth by increasing the number of ribosomes.

Exercise (Size distribution) *If the cells in balanced growth grow to a length (size) s_d and then divide precisely in half, then each cell has a size s where $(s_d/2) \leq s \leq s_d$. What is the probability distribution $P(s)$ for the size s in the population?*

Exercise (Age distribution) *If each cell in balanced growth grows for an interval τ_d and then divides, then each cell has an age τ where $0 \leq \tau \leq \tau_d$. What is the probability distribution $P(\tau)$ for the age τ? What is the average age τ of the cells in the culture? (The answer is not $\tau/2$.)*

---

[6] Note that if both rates follow Arrhenius laws, $k_r \propto \exp(-\Delta E_r/k_B T)$ and $k_t \propto \exp(-\Delta E_t/k_B T)$, then λ will also follow an Arrhenius law (equation (2.5)) with the average activation energy $\Delta E = (1/2)(\Delta E_r + \Delta E_t)$.

2.5 Partitioning of resources

As discussed above, when λ is varied by changing the nutrient availability, for a given bacterial strain the ribosome/protein ratio $r = R/P$ in (2.14) grows in proportion to λ. Because the ribosome is assembled from both RNA and protein, this ratio is an indicator of the fraction of cellular protein that is allocated toward making new protein. However, one can also vary λ by keeping the nutrient condition constant and varying the rate of translation—the efficiency of the ribosomes. One way to accomplish this is to add very small quantities of an antibiotic such as chloramphenicol, which impedes the synthesis of protein by the ribosome. Thus one may vary λ either by varying nutrient availability at fixed ribosome efficiency, or by varying the ribosome efficiency at fixed nutrient conditions. Observing how r changes in these different situations can give insight into how the cell allocates its protein resources.

Figure 2.1(A) summarizes the data from a careful study [8] of different *E. coli* strains growing under different nutrient and antibiotic conditions. Growth rate was manipulated by both the nutrient and ribosomal methods. The two methods generate very different linear relationships between λ and the RNA/protein ratio r. When the efficiency of the ribosomes is held constant but nutrient availability is altered, r follows

$$r = r_0 + \frac{\lambda}{\kappa_t}, \tag{2.15}$$

where r_0 and κ_t are constants. This is the line with positive slope in figure 2.1(A).

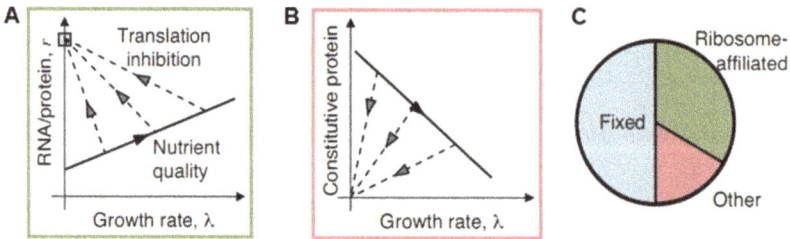

Figure 2.1. The cell's allocation of protein to different functions changes with its growth rate λ. (A) Because the ribosome is made from RNA and protein, the cell's ratio of RNA to total protein reveals the share of protein that is dedicated toward protein synthesis. The RNA/protein ratio depends on the growth rate λ. The growth rate of a given *E. coli* strain can be manipulated by adjusting nutrient availability, giving the solid line. Alternatively λ can be manipulated (at a fixed nutrient quality) by altering the translation efficiency, either using a different strain or supplying an antibiotic, giving the dashed lines. In either case, the relationship between RNA/protein ratio and λ is linear. (B) If a protein is constitutively expressed (and not part of the ribosome), its proportion within the total protein mass shows a linear behavior that is opposite to that in (A). When more protein resources are invested in the ribosomes, less are available for making the other protein. (C) Such data imply that certain functions receive a fixed allocation (shown in blue) of the cell's protein, while ribosomal (green) and other (pink) proteins split the remainder and receive λ-dependent shares respectively. Reprinted from [7]. Copyright 2011 with permission from Elsevier.

However, when the nutrient is held constant but the ribosome function is varied the relation is

$$r = r_{\max} - \frac{\lambda}{\kappa_n}, \tag{2.16}$$

where κ_n and $r_{\max}$ are constants. These are the negative sloping lines in figure 2.1A.

Better nutrient conditions reduce the slope $1/\kappa_n$, indicating that they increase κ_n. Better translational efficiency reduces the slope $1/\kappa_t$, indicating that it increases κ_t. Therefore κ_n was called the *nutritional* efficiency and κ_t was called the ribosomal (or *t*ranslational) efficiency [8].

The data also imply that the ratio r has an upper limit and a lower limit that are independent of κ_n and κ_t. This means there is an upper limit to the fraction of cellular protein that is tied up in ribosomes. This upper limit is substantially less than 100%, based on the maximum r in the data and the known proportionality between ribosomal protein and ribosomal RNA. Therefore, there must be another (non-ribosomal) fraction of the cellular protein that is being maximized when nutrient is scarce and minimized when nutrient is abundant. This is indicated in figure 2.1(B). In short, the cell's allocation of resources changes with its growth rate, in a way that depends on whether ribosome efficiency (via κ_t) or nutrient availability (via κ_n) is limiting to λ. The simple linear relationships between all these quantities motivated Scott *et al* to argue that cellular protein can be broadly divided into three fractions:

1. A fraction Φ_R of the total cellular protein is associated with ribosomes and is employed in the synthesis of additional protein. This fraction increases when nutrient is abundant.
2. A fraction Φ_P of the protein fosters growth but is not part of the ribosome. This fraction increases under poor nutrient conditions.
3. The remaining fraction Φ_Q of cellular protein is mostly independent of growth conditions.

The sum of the three fractions is fixed,

$$\Phi_R + \Phi_Q + \Phi_P = 1. \tag{2.17}$$

Based on the data, Scott *et al* argued that the fraction Φ_R is λ-dependent but not less than a minimum, Φ_R^0,

$$\Phi_R = \Phi_R^0 + \frac{\lambda}{k_t}, \tag{2.18}$$

where k_t (like κ_t) is a measure of ribosomal efficiency[7]. Rearranging (2.18) shows that λ is proportional to the degree to which Φ_R exceeds its baseline value,

$$\lambda = k_t(\Phi_R - \Phi_R^0). \tag{2.19}$$

Finally Scott *et al* proposed that the fraction Φ_P plays the role of bringing in the

---

[7] Because it relates ribosome content to the growth rate, this k_t is similar in spirit to the k_t that appears in the autocatalytic model. However it is not exactly the same.

nutrients used for protein synthesis. Growth depends on the flux of nutrients processed by Φ_P:

$$\lambda = \Phi_P k_n, \tag{2.20}$$

where k_n (like κ_n) is a measure of nutrient availability. In this way, the model suggests that the actively growing cell allocates its protein among different classes in such as way as to balance the import and processing of nutrient (equation (2.20)) with the use of that nutrient by the protein synthesis machinery (equation (2.19)).

The need to match the nutrient flux to the protein synthesis rate determines how much investment the protein must make in ribosomes as opposed to other machinery. Given the values of k_t and k_n, and with Φ_Q fixed, the three equations (2.17), (2.18), and (2.20) determine the three parameters Φ_R, λ, and Φ_P. A plot of Φ_R in terms of k_n and k_t, as shown in figure 2.2, shows for example how sharply Φ_R rises when translational efficiency falls and becomes limiting to growth. The result for λ is a familiar function

$$\lambda = k_t\left(1 - \Phi_Q - \Phi_R^0\right)\frac{k_n}{k_t + k_n}. \tag{2.21}$$

Because $k_t(1 - \Phi_Q - \Phi_R^0)$ is a constant, λ has the same functional dependence on the nutritional availability k_n as Monod's law, equation (2.4), which relates λ to the limiting nutrient concentration c. Although this model sets aside countless

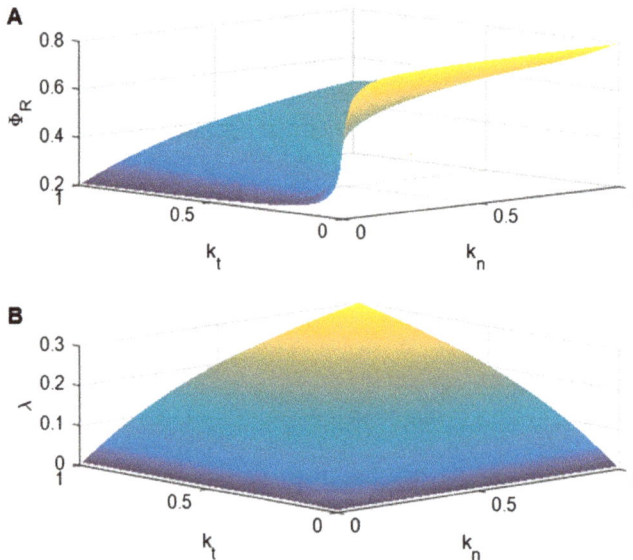

Figure 2.2. Sketch of the model of Scott *et al* [8] for the effect of translational and nutrient efficiency, k_t and k_n, on the growth rate and the allocation of protein. (A) The fraction Φ_R of cellular protein allocated to protein synthesis decreases if ribosomal function is more efficient, but increases under favorable nutrient conditions. (B) The growth rate predicted by (2.21), which is similar to the Monod law of (2.4), is maximal when both k_t and k_n are large. Balanced growth requires a match between the incoming flux of nutrients and their conversion to protein by the ribosomes.

biochemical details underlying bacterial growth, it does give an appealing physical perspective, in a simple top-down model, on how a cell allocates protein resources to optimize its rate of growth.

2.6 Individual cells in balanced growth

As mentioned above, when the growth rate is modified by adjusting the nutrient environment, the mass per cell and the RNA and DNA content per cell vary exponentially with λ. In addition, the average size of the cells varies according to the growth law

$$\langle v \rangle = A \exp B\lambda. \tag{2.22}$$

Here $\langle v \rangle$ is the average cell volume and A and B are constants.

One may wonder how this and other properties of balanced growth work at the level of the individual cell. Does the size of each cell increase exponentially in time? The fact that a cell doubles in size in a time interval τ_d does not require its length to increase exponentially in time. How much does the division time τ_d vary among individual cells? Surely the descendants of one parent cell will not all perpetually divide in synchrony, at precise intervals of τ_d. And does the cell rely on size or growth time to trigger division? Whatever the mechanism that triggers a division event, it should impose some limit to the variability in τ_d, or else after a few generations some cells by chance may become freakishly large or small.

Recent developments in microfluidics allow researchers to answer these questions by capturing and monitoring individual cells and tracking their growth and division over many generations. The red curve in figure 2.3 shows a recent test of the growth

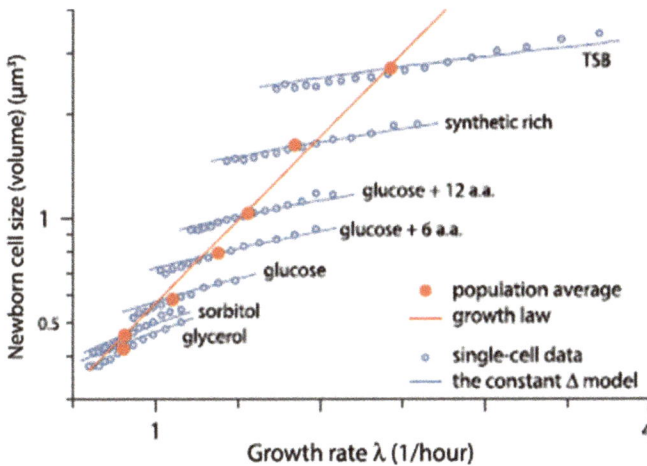

Figure 2.3. The dependence of growth rate on cell size in *E. coli*. Growth rate is modulated by changing the nutrient content of the medium. The population-averaged cell size increases exponentially with the average growth rate (red points and line), as in (2.22). However, under any particular nutrient condition, the size of an individual cell depends only weakly on that cell's own doubling time (blue circles). The circles in the figure represent binned results of many individual cell measurements. Reprinted from [9]. Copyright 2015 with permission from Elsevier.

law of (2.22) in *E. coli*. Consistent with (2.22), as growth rate is varied by changes in nutrient condition, the population-averaged volume of the newborn cells increases exponentially with the population growth rate λ. However, figure 2.3 was obtained by making a large number of precise microscopy measurements of the size and growth rates of individual cells. It reveals some puzzling details that are not captured by (2.22). While the population-averaged behavior does follow the red curve of the growth law, under any particular growth condition the growth rate λ of individual cells and the volume v_b of the newborn cells are subject to variability. This variability does not obey (2.22). Evidently the relation between λ and v_b for individual cells at a particular growth condition is not the same as the relation between population-averaged λ and v_b as growth conditions vary [9]. Single cells exhibit complexity in behavior that is not observable in a population.

The data in figure 2.3 relate to the basic question of how individual bacteria grow in time, and how they control the timing of division. To determine whether individual cells grow linearly or exponentially in time is an experimental challenge, as linear and exponential growth may not differ greatly over time intervals as short as τ_d. To monitor the growth of individual cells, one can trap cells in a microscopic chamber and observe their length $s(t)$ as they grow and divide. Many bacterial cells have cylindrical shape that simply elongates as they grow, without changing in width. The cell's length, volume, and visible area are therefore all suitable indicators of its size. Rather than measure the volume v of the cell one may measure just its length s in a microscopy image.

Such careful imaging of large numbers of cells has clarified that at least for some species, the length $s(t)$ of each cell truly does grow exponentially [10]. Figure 2.4 shows data from five *C. crescentus* cells, dividing over 300 generations. The area of

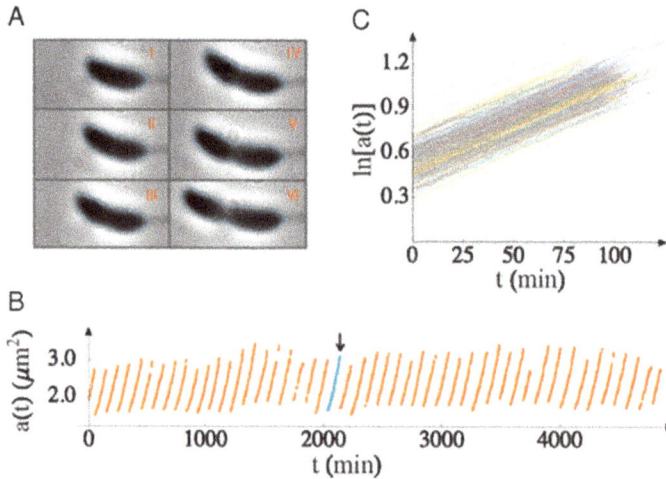

Figure 2.4. Repeated growth and division events of an individual *C. crescentus*. (A) Asymmetric division of a cell attached to a glass surface. (B) Time dependence of the imaged area a (proportional to length s) of the parent, as it divides repeatedly. The size at the instant of division is variable from one division event to another. (C) The size of the parent cell increases exponentially in time until division occurs, as indicated by the linear dependence of a on a semilogarithmic plot. Reproduced from [10] with permission.

the cells increases linearly on the logarithmic vertical scale of the figure, indicating that the length s of each cell grows exponentially in time,

$$s(t) = s_b \exp(\lambda t), \tag{2.23}$$

until division occurs at $t = \tau_d$. Here s_b is the length of the newborn cell.

But it is also clear from figures 2.3 and 2.4 that both τ_d and the length $s(\tau_d)$ (or equivalently the imaged area a) at the instant of division are variable within a population. A parent cell and its daughter may divide at slightly different intervals and sizes. Thus it is not true that every cell grows for precisely the same period τ_d before dividing: τ_d has a probability distribution. Although at the population level τ_d and λ describe the same property of doubling rate, at the level of the individual cell they describe different quantities. Each cell's λ is defined by (2.23) and by the slope of an individual curve in figure 2.4(C). Its τ_d is defined by the duration of that curve.

Of course λ and τ_d are correlated. Figure 2.5 shows for C. crescentus that, despite some scatter, the two quantities tend to vary inversely: $\lambda \simeq r/\tau_d$ with the same constant $r = 0.565$ applying over a temperature range 17 °C–34 °C. The average size at division is then $s_d = s_b \exp(\lambda\tau_d) = s_b \exp(r)$, or $s_d/s_b \simeq 1.76$. Thus, while the growth rate and the division time of cells can both fluctuate rather significantly, these cells tend to divide when their length or volume has increased 76% from birth.

Studies of E. coli and B. subtilis, which divide symmetrically, also suggest that it is the size of the growing cell that triggers its division. Comparing the size at division (s_d) to the size at birth s_b shows that cells consistently divide when their size has increased by a fixed increment Δ beyond s_b [9]. That is, a cell divides when it reaches a length

$$s_d = s_b + \Delta, \tag{2.24}$$

creating two daughters of size $s_d/2$. Δ is very nearly constant under any growth condition although, as figure 2.6 shows, Δ and s_b vary considerably with the growth

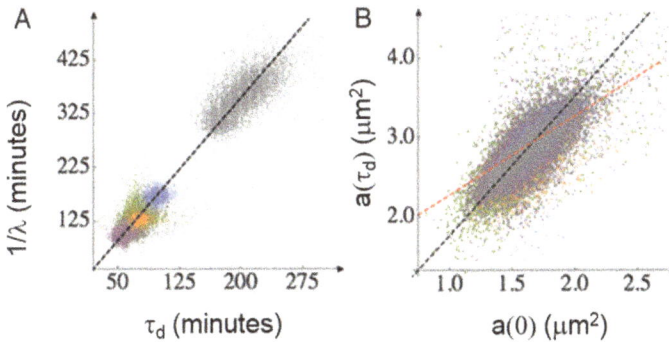

Figure 2.5. For division of individual C. crescentus, the inverse growth rate $1/\lambda$ is proportional to the division time τ_d. (A) Cell division events studied at different temperatures (color coded) ranging from 17 °C (gray) to 34 °C (purple) fall on the same proportional curve, indicating that τ_d and λ vary with environmental conditions but generally $\lambda\tau_d \simeq 0.565$. (B) The cell area at the instant of division (vertical axis) is compared to the area of the same cell at birth (horizontal), indicating that division occurs when the cell size has increased by a factor $\exp(0.565) = 1.76$ (black line). Reproduced from [10] with permission.

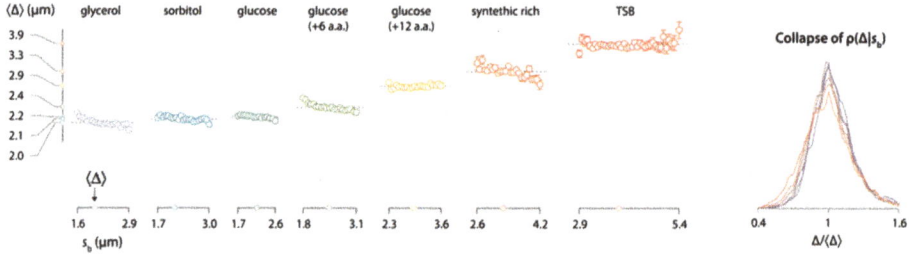

Figure 2.6. The length s_b of a newborn *E. coli* cell varies within a population, and also depends on the growth conditions such as the nutrients in the growth medium. However, for a given growth condition the difference Δ between the length of the cell at division and its newborn length s_b is nearly constant. For each nutrient source as indicated, the Δ values at which division occurs are rather tightly clustered around a mean value $\langle\Delta\rangle$, with a probability distribution for $\Delta/\langle\Delta\rangle$ that is similar under all conditions. Similar results were observed with *B. subtilis*. Reprinted from [9]. Copyright 2015 with permission from Elsevier.

conditions. More precisely, Δ has its own probability distribution (like τ_d above) with a well-defined average $\langle\Delta\rangle$ under any particular growth condition.

One of the wonders of (2.24) is that it is a straightforward way of keeping the average cell size constant, even if the division mechanism is slightly unreliable and Δ fluctuates from one division to the next. If each cell partitions into two equal daughters, then the size of a newborn daughter is $\langle s_b\rangle = \Delta$. If by chance Δ is unusually large or small at one division, the daughter will have an unusually large or small s_b, denoted by $\langle s_b\rangle + \delta s_b$. This cell will later divide to create two daughter cells of size $s_b = \frac{1}{2}\langle s_b\rangle + \frac{1}{2}\delta s_b + \frac{1}{2}\Delta$. After n generations the daughter cells will be of size $s_b = \langle s_b\rangle + \frac{1}{n}\delta s_b$, as long as the fluctuations in Δ are uncorrelated. In this way the effects of fluctuations and occasional outlier values of Δ are diluted away over the generations and the average newborn cell size tends toward $s_b = \langle\Delta\rangle$. Sattar *et al* observed exactly this behavior [9].

The origins of these apparent mechanisms in cell division, as well as their variability, are still not understood. Still they show some curious properties. For example, the division time τ_d for *C. crescentus* has a certain stability. If $\langle\tau_d\rangle$ is the average division time, this average changes with the growth condition (nutrients, temperature). But if the division time τ_d is normalized to its average at any condition $x = \tau_d/\langle\tau_d\rangle$, the probability distribution $P(x)$ for the normalized division time is the same over a range of growth conditions [10]. This is somewhat surprising as the growth rate of *C. crescentus* varies about sixfold between 37°C and 14 °C, following the Arrhenius law of (2.5) with an activation energy $\Delta E \simeq 54$ kJ mol^{-1}. For the symmetrically dividing species the probability distribution for the normalized $\Delta/\langle\Delta\rangle$ is likewise similar under all growth conditions [9]. So while there is noise in cell division, that noise shows intriguing, simplifying aspects of universality that were not necessarily anticipated on theoretical grounds. These findings should be useful in narrowing down possible physical models for the regulation of balanced growth.

Therefore while bacterial growth has long been a rich topic for study at the bulk level, single-cell studies demonstrate there is much more to see in the kinetics and statistics of cell growth. Cell-to-cell variability will play a pronounced role in gene regulation and signaling, to be discussed later.

References

[1] Monod J 1949 The growth of bacterial cultures *Ann. Rev. Microbiol.* **3** 371–94

[2] Ratkowsky D A, Olley J, McMeekin T A and Ball A 1982 Relationship between temperature and growth rate of bacterial cultures *J. Bacteriol.* **149** 1–5 PMCID: PMC216584

[3] Verhulst P F 1838 Notice sur la loi que la population suit dans son accroissement *Correspond. Math. Phys.* **10** 113–21

[4] Hagen S J 2010 Exponential growth of bacteria: constant multiplication through division *Am. J. Phys.* **78** 1290–6

[5] Campbell A 1957 Synchronization of cell division *Bacteriol. Rev.* **21** 263–72

[6] Schaechter M, Maaloe O and Kjeldgaard N O 1958 Dependency on medium and temperature of cell size and chemical composition during balanced growth of Salmonella typhimurium *J. Gen. Microbiol.* **19** 592–606

[7] Scott M and Hwa T 2011 Bacterial growth laws and their applications *Curr. Opin. Biotechnol.* **22** 559–65

[8] Scott M, Gunderson C W, Mateescu E M, Zhang Z and Hwa T 2010 Interdependence of cell growth and gene expression: origins and consequences *Science* **330** 1099–102

[9] Taheri-Araghi S, Bradde S, Sauls J T, Hill N S, Levin P A, Paulsson J, Vergassola M and Jun S 2014 Cell-size control and homeostasis in bacteria *Curr. Biol.* **25** 385–91

[10] Iyer-Biswas S, Wright C S, Henry J T, Lo K, Burov S, Lin Y, Crooks G E, Crosson S, Dinner A R and Scherer N F 2014 Scaling laws governing stochastic growth and division of single bacterial cells *Proc. Natl Acad. Sci.* **111** 15912–7

The Physical Microbe
An introduction to noise, control, and communication in the prokaryotic cell
Stephen J Hagen

Chapter 3

Gene regulatory networks

The cell must perform a variety of tasks in order to gather and process nutrients, grow, locomote, replicate, and respond to its environment. Its survival depends on its ability to manufacture and maintain a biomolecular toolkit that consists of thousands of different peptides, proteins, and RNAs, as well as small molecules and larger structures such as membranes and ribosomes. The toolkit and the instructions for regulating its production and maintenance are encoded in the genome of the organism. The task of using that information to control the cell's behavior is performed by networks of molecules that interact and react, gathering and processing information about the internal state of the cell as well as the extracellular environment. The literature contains many illustrations of these complex biochemical networks. Some of these networks can be described as enzyme pathways, describing a set of reactions associated with the biochemical task performed by a particular enzyme. Others are metabolic pathways, or linked reactions that generate a flux of synthesis or breakdown of a particular type of molecule.

We will focus on gene regulatory pathways and, to a lesser extent, signal transduction pathways. A gene regulatory pathway is the sequence of interaction events that controls production of a particular gene product. The product may be a protein, a peptide, or an RNA. As a gene regulatory pathway may involve activation and expression of multiple genes, the flow of information through the pathway may require time scales of minutes or hours.

By contrast a signal transduction pathway is the cascade of biochemical reactions that follows, most typically, the detection of a signal molecule by a receptor molecule. These pathways often involve rapid chemistry such as phosphorylation/dephosphorylation (discussed below). Therefore they can potentially process information much more quickly than gene regulatory pathways, in seconds or faster. When bacteria swim actively toward food, for example, a signal transduction pathway controls their movement.

doi:10.1088/978-1-6817-4529-9ch3

Both of these types of pathways often consist of multiple, interacting modules that signal to, control, activate, and deactivate each other and other related pathways. As a result they can exhibit complex time-dependent behaviors such as feedback, switching, or oscillation. In addition, the physical phenomena of noise or stochasticity, as well as other environmental cues such as signals received from other cells, can shape the output. One can gain a physical perspective on these phenomena by modeling the production of a single protein from a single regulated gene. Therefore this chapter begins with a brief review of genes and the fundamentals of transcription and translation. It then examines some of the feedback and dynamical phenomena that are possible due to gene regulation mechanisms. This will then lead to consideration of stochasticity in gene expression. Later we will look at an example, the bacterium *Bacillus subtilis*, in which nonlinear dynamics of regulation combine with stochasticity and intercellular signaling to produce truly sophisticated bacterial decision-making.

3.1 Transcription and translation

For nearly every job that needs to be done, the cell makes a protein[1]. As the protein is a polypeptide, or polymer of amino acids[2], its physical and chemical properties are determined by the sequence of amino acids that occurs along the polypeptide chain. The sequence for each protein is encoded as information—a gene[3]—in the organism's DNA. This information is not stored in the form of amino acids, however, but as a coded sequence of nucleotide bases (A, T, C, or G).

More formally, a gene is a sequence of DNA that encodes a functional molecular product. The readout of the gene to the functional product involves the transcription of the DNA nucleotide sequence into an RNA nucleotide sequence; if the gene encodes a protein the transcription is then followed by the translation of that messenger RNA into the language of amino acids. If the gene product is an RNA, it may be used structurally, perhaps as a portion of a ribosome, or as a regulatory element that interacts with a particular mRNA[4].

Several different types of information are physically encoded in the gene. Of the two DNA strands, one strand carries the nucleotide sequence that acts as

---

[1] The term protein and enzyme are often used synonymously, but an enzyme is a biomolecule—most often a protein—that catalyzes a reaction. Catalase is a protein, but it is also an enzyme as it breaks down hydrogen peroxide to water and oxygen.

[2] Recall that virtually all proteins are assembled from the same twenty amino acids (glycine, alanine, tryptophan, tyrosine, and so on), taken in different combinations. The amino acids are condensed into a polymeric chain but retain their individual side chains, which add the distinctive physical and chemical properties of the protein.

[3] So as not to confuse a gene with its protein (or RNA) product, the name of the gene is usually set in italics, with the first letter in lowercase. The gene product is set in upright (roman) type, with the first letter in uppercase. For example, the product of the gene *perR* is PerR, a protein involved in regulating response to stress.

[4] mRNA = messenger RNA.

template for the mRNA. The other, complementary DNA strand is called the coding strand as it has the same nucleotide sequence as the mRNA. The information on the template strand is read by an enzyme, RNA polymerase. The polymerase travels along the template, from the transcription start site to the terminator site, synthesizing the mRNA as a continuous strand. Immediately upstream of the start site is a region of the gene known as the promoter. This is the region that the RNA polymerase recognizes and binds to, in order to begin transcription. The promoter indicates which DNA strand needs to be transcribed and it defines the direction of transcription. The RNA polymerase recognizes the promoter by the presence of two specific short nucleotide sequences that are 10 nucleotides and 35 nucleotides, respectively, upstream of (prior to) the transcription start site[5]. Thus transcription consists of the three steps of (1) initiation, in which the RNA polymerase recognizes the promoter and begins transcribing the template, (2) elongation, in which the RNA polymerase travels along the template strand making its mRNA copy, and (3) termination, where the RNA polymerase recognizes a terminator sequence on the template and dissociates, halting transcription.

The cell has many ways to regulate transcription and translation in order to control the amount of active protein. The cell can apply transcriptional regulation, meaning that it modulates the rate of the initial transcription (RNA message production) step, enhancing or suppressing this readout through molecules that physically interact with the particular gene and the polymerase. The protein can be translationally controlled, meaning that its RNA message is transcribed but the mRNA is intercepted (such as by other RNAs) before being translated by the ribosome. In addition, many proteins are modified or regulated post-translationally, after the polypeptide chain has been synthesized. The chain may be cleaved or truncated, separate chains may be combined, covalently or otherwise, into one assembly, and additional chemical groups (cofactors) or other chemical modifications may be added. Such modifications affect the protein's function. One of most common post-translational modifications is phosphorylation, which is the covalent attachment of the phosphoryl (PO_3^{2-}) group to one of the protein's side chains. It is a simple, reversible modification that can immediately activate or deactivate an enzyme. Often in a signal transduction pathway, a series of proteins will hand off phosphoryl groups to each other, forming a so-called phosphorylation cascade. In addition enzyme activity can be altered by the binding of a ligand or effector molecule. The end product of an enzymatic pathway may for example interact with an enzyme earlier in the pathway, acting as a feedback signal to regulate flux through the pathway.

---

[5] If a gene is named *xyz*, its promoter sequence will often be referred to as P*xyz*. This is useful for example when the P*xyz* sequence is synthetically inserted upstream of another transcription start site, such as one encoding a fluorescent protein like green fluorescent protein (GFP). This is how one creates a fluorescent reporter, a synthetic gene that will express fluorescent protein when the cell attempts to express *xyz*.

3.2 Representation of networks and pathways

In studying gene regulation one sees many diagrams of biochemical networks. Most of these networks are drawn in similar ways, although they are not nearly as standardized as road maps or electrical circuit diagrams. As a general rule, if molecule or component A has a positive effect on the action or abundance of component B, this is usually indicated with an arrow: $A \rightarrow B$. A negative effect of A on B is usually indicated with a transverse bar: $A \dashv B$. Unfortunately this notation is not very specific as numerous microscopic mechanisms are consistent with these representations. For example an arrow from A to B may indicate that A transforms to B, or that A promotes (by an unspecified mechanism) the production of B. In the latter case, the functionality of B may or may not actually require A. In any case, the full mechanism may be so convoluted (or so poorly understood) that even a good drawing may not shed much light. The problem of interpretation worsens as increasing numbers of molecular species and interactions are identified, leading to more tortuous diagrams. Fortunately some authors have proposed standards to clarify the visual presentation of interaction networks. Figure 3.1 shows a few symbols from a scheme proposed by Pirson *et al* [1], which we will attempt to follow. As is typical, an arrow generally denotes a positive or enhancing interaction, while a bar denotes some form of inhibition or repression. Chemical transformation is distinguished from activation or transcriptional control (described below). When transcription needs to be in the picture, a bent arrow indicates the transcription start site and direction. Diagrams under this scheme are still not unique for any given model, but the scheme does distinguish usefully between common interactions. Figure 3.2 gives some examples of the system in use.

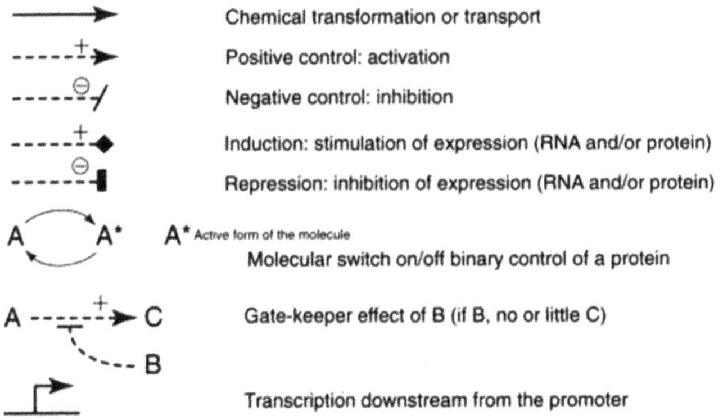

Figure 3.1. A scheme for drawing biological regulatory networks, proposed by Pirson *et al* [1]. It aims to distinguish graphically between several possible mechanisms by which the action of molecule A can impact the action of B. A can transform to B, or it can activate or inhibit the function of B, such as by chemical modification of B. Alternatively A can stimulate (or inhibit) the expression of B. Reprinted from [1]. Copyright 2000 with permission from Elsevier.

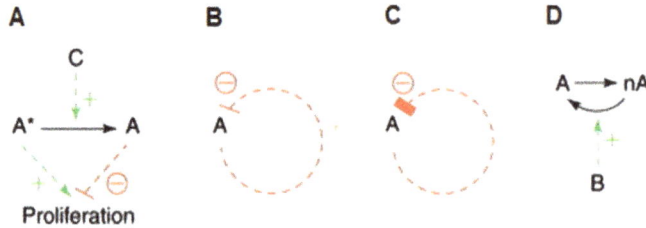

Figure 3.2. Some examples of regulatory interactions in the Pirson *et al* scheme. (A) The activated form of molecule *A* (which promotes proliferation) transforms to an inactivated form that inhibits proliferation. *C* stimulates this transformation. (B) The molecule *A* inactivates itself. (C) *A* is a regulatory molecule that inhibits its own expression. (D) *A* polymerizes reversibly, forming a complex of size *n*. *B* causes the complex to unravel. Reprinted from [1]. Copyright 2000 with permission from Elsevier.

3.3 Gene regulation basics

As discussed above the cell can adjust the activity of functioning enzymes through mechanisms such as phosphorylation or binding of ligands. In addition the cell actively regulates the expression of genes: it modulates the processes of transcription and translation that generate the physical, functional products—protein or RNA—of those genes. The regulation of expression is a critically important enterprise that raises a number of physical issues related to sensing, information processing, dynamics, and noise. Many genes encode products used solely for regulation. Those genes whose products are not employed for regulation are called structural genes.

There are many mechanisms by which bacteria regulate gene expression, including RNA-driven mechanisms that affect transcription as well as translation. Regulation in archaea is mostly similar to bacteria. However, eukarya utilize additional layers of control such as DNA methylation, chromatin structure, and the splicing and editing of mRNA prior to translation. Here we only review some of the basic modes of bacterial transcriptional control that will arise in later discussions[6].

The simplest gene regulation is none at all. Some genes are continuously expressed and not otherwise regulated. These are called constitutive genes. More typically, however, the cell modulates expression of a gene up or down to meet current need. One may then refer to genes being switched on or off, although a gene is rarely completely switched off. Regulation is usually a matter of degree, with the cell tuning expression up or down over a dynamic range. Regulated genes are said to be either inducible or repressible. If a gene's transcription is usually slow or infrequent but the cell has a method for increasing this activity, the gene is said to be inducible. If a gene is usually more active until the cell applies a mechanism to suppress it, the gene is said to be repressible. Transcription from inducible and repressible genes is usually controlled by proteins called transcription factors[7], which

---

[6] The reader should look up additional details in a microbiology textbook such as [2].

[7] Expression of the gene can also be modulated by elements that interact after transcription has occurred, such as by regulatory RNAs interacting with the mRNA transcript before translation.

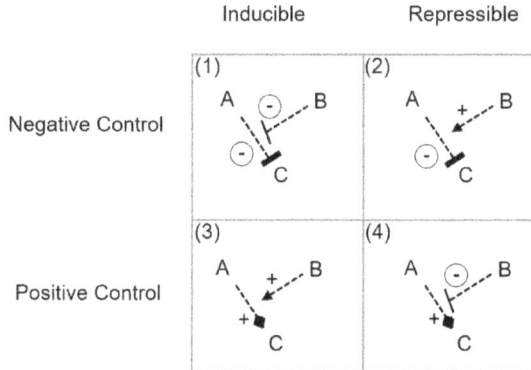

Figure 3.3. Mechanisms of gene regulation by transcription factors. A transcription factor A controls transcription of gene C, but the action of A is modulated by the effector molecule B. Depending on whether the effect of A on C is positive or negative, and depending on whether the effect of B is to induce or repress transcription, B is called (1) an inducer, (2) a co-repressor, (3) an inducer, (4) an inhibitor of C. The arrow notation of figure 3.1 is used.

interact either with the promoter or the RNA polymerase itself. To make things slightly confusing, transcription factors can act on both inducible and repressible genes in either of two ways, as follows:

- A transcription factor may interact with a specific sequence of DNA near the transcription start site, so as to block or inhibit the function of the RNA polymerase. In this case the gene is said to be under negative transcriptional control. The transcription factor is then called a repressor and the interaction site is called an operator.
- The transcription factor may instead have the opposite behavior, interacting with the DNA sequence in such as way as to promote transcription. The gene is then said to be under positive transcriptional control, and the transcription factor is called an activator.

Therefore there are four main possibilities for transcriptional control: the gene may be inducible or repressible, and this induction or repression may be under either positive or negative control.

The cell may regulate transcription of a gene by changing the concentration of a relevant transcription factor. It may also use a small molecule to activate or deactivate the transcription factor. By binding to the transcription factor, the ligand may positively or negatively modulate its affinity for the binding site. The effector molecule is called an corepressor if it enhances the action of a repressor, an inducer if it alleviates that repression or enhances the binding of an activator, or an inhibitor if it reduces the binding of an activator.

These different possibilities generate the distinct cases that are shown in figure 3.3. There are famous examples of each case. Perhaps the most famous example of a regulated gene in *Escherichia coli* is that of the *lac* operon[8], which encodes enzymes to import and break down lactose sugar. The operon is under negative control but is inducible by allolactose, an indicator of the presence of lactose. Consequently the *lac* operon is transcribed at only minimal levels unless lactose is present for the enzymes to act upon, in which case transcription ramps up.

Information about the extracellular environment is often fed into the control of transcription through dedicated two-protein systems known as two-component signal transduction systems (TCSTSs). In a TCSTS, a receptor protein that resides in the cell membrane detects an extracellular signal or cue. The receptor[9] responds by phosphorylating a second, intracellular protein known as the response regulator. The phosphorylated response regulator then acts as a transcriptional activator or repressor for a regulated gene.

3.4 Deterministic models for gene regulation

Having reviewed some of the biology, we can now study some physical behaviors of regulated genes. We start with simple deterministic models for induction and repression. These will show how nonlinear behaviors such as bistability can arise in gene expression. The models are deterministic in the sense that, given a set of initial conditions (initial concentrations of the various molecular species), they completely predict the subsequent time evolution of the protein concentration. However, while they indicate what to expect on average from a population of cells, deterministic models do not capture the effects of molecular-level fluctuations in the underlying chemical and physical processes. Such effects drive cell-to-cell variability that is very apparent in real bacteria. To understand how molecular-level fluctuations affect regulatory pathways, we will later revisit regulation from a stochastic perspective[10].

Starting with an unregulated gene, and leaving RNA out of the picture, one can create a deterministic model for the production and degradation of a protein A by writing a differential equation for the time dependence of n_A, the number of copies of A in a cell, given that A is synthesized at a constant rate k (molecules/s in one cell). The model should account for the fact that A does not accumulate without limit. Some amount of A is lost continuously, either because it is actively degraded or simply because of dilution: half of the copies of A are lost with every cell division. For stable proteins dilution is the most important mechanism of loss. In any case, loss of A is characterized by a rate γ (fraction lost per unit time). Then n_A evolves according to

---

[8] An operon is a section of DNA containing several structural genes, which are transcribed as a unit. The *lac* operon contains three structural genes, *lacZ*, *lacY*, and *lacA*.

[9] The receptor is known as the sensor kinase of the TCSTS, as a kinase is an enzyme that phosphorylates another protein.

[10] Stochastic means randomly determined, characterized by a random variable that may be analyzed statistically but not predicted precisely.

$$\frac{dn_A}{dt} = k - \gamma n_A. \tag{3.1}$$

Figure 3.4(A) summarizes the model. If A is lost through dilution, then $\gamma = \log(2)/\tau_d$ (equation (2.1)). Equation (3.1) indicates that n_A will approach a steady state $(dn_A/dt = 0)$ where $n_A = k/\gamma$. Kinetically this occurs as an exponential relaxation, characterized by the rate γ,

$$n_A = \frac{k}{\gamma}(1 - e^{-\gamma t}). \tag{3.2}$$

To make a more sophisticated model that takes into account both transcription and translation, one may replace k with a pair of production rates as in figure 3.4(B): k_R (mRNA copies produced, per unit time) is the rate of transcription of a, and k_P (copies of A produced per unit time, per mRNA) is the rate of translation. There will also be two degradation rates: γ_R is the degradation rate of mRNA and γ_P is the degradation rate of the protein. If n_R is the number of copies of the mRNA, then

$$\frac{dn_A}{dt} = k_P n_R - \gamma_P n_A, \tag{3.3}$$

$$\frac{dn_R}{dt} = k_R - \gamma_R n_R. \tag{3.4}$$

The system will relax exponentially toward its steady state $(dn_A/dt = dn_R/dt = 0)$, where

$$n_R = \frac{k_R}{\gamma_R}, \tag{3.5}$$

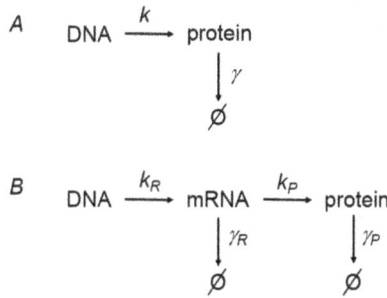

Figure 3.4. (A) A simple deterministic model for constitutive expression of a protein. Transcription and translation are combined into a single process with a rate k (protein molecules/time). The protein is degraded at a rate γ. (B) A slightly more realistic model includes transcription, with mRNA produced at a rate k_R (molecules/time), and translation, with protein produced at a rate k_P (molecules/time/mRNA). mRNA and protein are degraded or diluted at rates γ_R and γ_P, respectively. The models do not have very interesting deterministic behavior, but they are just detailed enough to be useful in understanding how stochasticity affects gene expression in experiments such as in [3, 4].

$$n_A = \frac{k_P k_R}{\gamma_P \gamma_R}. \tag{3.6}$$

Equation (3.6) suggests that different combinations of transcription and translation rates could lead to the same steady state n_A. If k_P/γ_P is large while k_R/γ_R is small, the number of mRNA copies present at equilibrium will be small, but each mRNA will generate many copies of the protein. Conversely if k_R/γ_R is large while k_P/γ_P is small, there will be many mRNA copies but each will generate relatively few protein copies. Because both scenarios can lead to the same steady state n_A, the deterministic model conceals an important difference between the two cases: when production and degradation are controlled by random underlying processes, the cell-to-cell variability in protein copy number may depend a great deal on whether the average n_R is large or small, even if the average copy number n_A is the same.

Now consider an actively regulated gene. Returning to the simpler (single production rate k) model of (3.1), A may be subject to transcriptional regulation. If molecule B is a transcriptional repressor for gene a, it suppresses the transcription rate k by binding to the operator site. That binding is a chemical interaction, for which the probability that B is absent is $K/(n_B + K)$. Here n_B is the number of copies of B and K is the dissociation constant of the interaction. Then if B is a transcriptional repressor for a, the repressing action of n_B can be described by

$$\frac{dn_A}{dt} = k\frac{K}{n_B + K} - \gamma n_A. \tag{3.7}$$

Conversely if B is a transcriptional activator, then the full transcription rate requires binding, which has probability $n_B/(n_B + K)$. Then the expression is

$$\frac{dn_A}{dt} = k\frac{n_B}{n_B + K} - \gamma n_A. \tag{3.8}$$

In some cases the transcription factor forms a dimer when it interacts with the promoter, or it requires other, multi-particle interactions to have its effect. The need to form these multi-particle interactions makes the transcription factor's effect on the gene more sensitive to its intracellular concentration. The function of an activator B is then better described by the so-called Hill function,

$$\frac{dn_A}{dt} = k\frac{n_B^m}{n_B^m + K^m} - \gamma n_A. \tag{3.9}$$

Here m is known as the Hill coefficient. If B is a repressor,

$$\frac{dn_A}{dt} = k\frac{K^m}{n_B^m + K^m} - \gamma n_A. \tag{3.10}$$

Higher values of m give a more strongly nonlinear, all-or-nothing[11] character to the switching between high and low transcription rates, as a function of n_B.

---

[11] The Hill function $f(x) = x^m/(x^m + K^m)$ develops a stronger sigmoidal or step-like dependence on x as m increases. This sigmoidal behavior is referred to as co-operativity.

There can be interesting consequences when A acts as its own transcriptional activator (or repressor), as is sometimes the case. For example, if A activates its own transcription the deterministic model is

$$\frac{dn_A}{dt} = k\frac{n_A^m}{n_A^m + K^m} - \gamma n_A. \tag{3.11}$$

First notice the behavior in the case $m = 1$. n_A will have two possible steady states, defined as states for which $dn_A/dt = 0$. There is an inactive state at $n_A = 0$ and an activated state at $n_A = k/\gamma - K$. (Of course $n_A \geqslant 0$ so if K is very large there can only be one steady state, at $n_A = 0$.) The fact that $dn_A/dt = 0$ in both states does not, however, mean that n_A is stable at both values. Stability of these states depends on the values of k and γ. Suppose that n_A is very small relative to K. Then $dn_A/dt \simeq (k/K - \gamma)n_A$. If production is brisk ($k/K > \gamma$) the net protein accumulation will be positive, so the system will evolve toward the activated state, which must then be the stable equilibrium. But if production is feeble ($k/K < \gamma$) then $dn_A/dt < 0$ and the system will tend toward $n_A = 0$, which will be the stable equilibrium. This model describes a gene that can keep itself switched on or off.

If the Hill coefficient is larger, $m = 2$, then an autoregulated gene is capable of more interesting behavior. An $m = 2$ system has three possible steady states, as can be seen by applying the condition $dn_A/dt = 0$ to (3.11).

$$n_A\left(\gamma n_A^2 - kn_A + \gamma K^2\right) = 0. \tag{3.12}$$

One solution is the inactive state, $n_A = A_0 = 0$. Two other solutions are $n_A = A_\pm$ where

$$A_\pm = \frac{k \pm \sqrt{k^2 - 4\gamma^2 K^2}}{2\gamma}. \tag{3.13}$$

If production is too slow (k too small) then $A_\pm$ are unphysical (complex) solutions, so only the inactive state A_0 is possible. If production is fast relative to degradation, $k^2 > 4\gamma^2 K^2$, then $A_\pm$ are both real. Then the system has an inactivated state A_0 and two activated states. Then as figure 3.5 shows, the A_0 and A_+ states will be stable: if n_A departs slightly from either A_0 and A_+, it is driven back toward that state. Thus the gene for A exhibits a bistability, where it can persist in either the inactivated or activated state. By contrast A_- is an unstable steady state, as small departures from A_- tend to grow over time.

Because the bistable system has two stable equilibrium states, the actual state of n_A will depend on the prior history of the system. In addition, the system can be driven between the two stable states by coupling to another gene product, or by stochasticity in the expression of A.

Nonlinearity and, in particular, bistability in gene expression allow a variety of complex and useful behaviors in networks containing multiple interacting elements. For example, one can extend (3.11) to the case of mutual regulation of two gene products: if A and B mutually inhibit each other's production as in figure 3.6 (here with rates k_A, k_B and dissociation constants K_A, K_B),

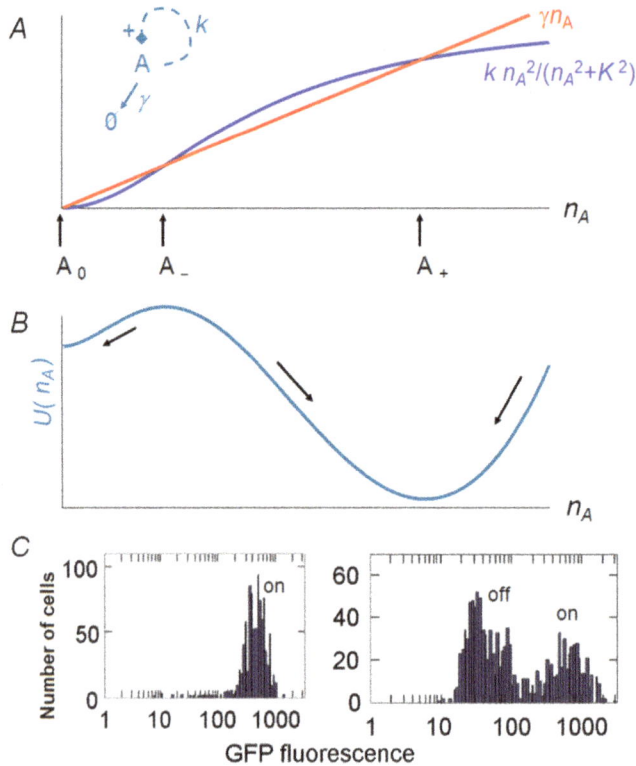

Figure 3.5. Multiple steady states of the positively autoregulated gene product A. (A) The production (rate k) and degradation (rate γ) terms for A are sketched in blue and red, respectively, based on (3.11) with $m = 2$. The system has steady states when $dn_A/dt = 0$, which occurs at the three values of A where the curves intersect. Because the degradation curve has the greater slope near A_0 and A_+, the net dn_A/dt is negative (or positive) if n_A is slightly larger (or smaller, respectively) than either of these two values. Consequently A_0 and A_+ are stable equilibrium points. By contrast, the production curve has greater slope near A_-, so the net dn_A/dt tends to pull the system away from A_-, making A_- an unstable steady state. The system is said to be bistable, with an inactive stable fixed point at A_0 and an active one at A_+. (B) Intuitively the evolution of $n_A(t)$ in (3.11) resembles the trajectory of a particle executing heavily damped motion on a hypothetical energy surface $U(n_A)$, which is defined so that $-dU/dn_A \propto dn_A/dt$. The particle is pulled toward the minima in U which are the attractors at A_0 and A_+. These stable fixed points are separated by the maximum in U, which prevents the system from transiting between the inactivated and activated states in the absence of fluctuations or external influences. (C) Bistability of this type occurs in the expression of the gene *comX* in *Streptococcus mutans*. The activity of *comX* is controlled by a positive feedback loop, where the strength of the feedback depends on environmental conditions. The figure shows histograms of *comX* activity in a population of cells, as observed through expression of a GFP reporter for P*comX*, under two types of environmental conditions. The feedback circuit can operate in a monostable fashion (left), with one stable state (with ComX production switched on), or it can operate in bistable mode (right) where two stable states exist as in (A) and (B). In the bistable case the population splits into two subpopulations with *comX* on and off, respectively. Adapted from [5].

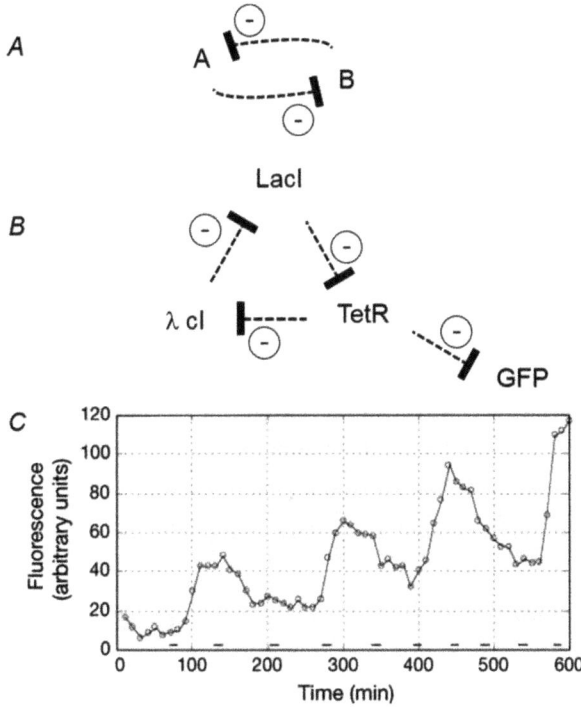

Figure 3.6. (A) A toggle switch consisting of mutually repressing gene products A and B. (B) Schematic of the repressilator [6], a synthetic gene network in which the three gene products TetR, LacI, and λ cI mutually repress each other's production. The state of the system is read out through the fluorescence of GFP, which is under control of TetR. (C) Oscillations in the repressilator implanted in *E. coli*. In an individual cell the GFP fluorescence, which reports on TetR, oscillates with a period of about 150 min, longer than the cell division time, which is approximately 1 h. Panel (C) reprinted by permission from Macmillan Publishers Ltd: [6], copyright 2000.

$$\frac{dn_A}{dt} = k_A \frac{K_A^2}{n_B^2 + K_A^2} - \gamma_A n_A, \tag{3.14}$$

$$\frac{dn_B}{dt} = k_B \frac{K_B^2}{n_A^2 + K_B^2} - \gamma_B n_B, \tag{3.15}$$

then activation of A inhibits B, and vice versa. Either A or B, but not both, can be activated at one time. Nevertheless (subject to the parameter values) the system is equally at equilibrium regardless of whether A or B is switched on. A system with this property is referred to as a toggle switch.

If three gene products A, B, and C interact in mutually repressing fashion, then some values of the interaction parameters may allow no stable steady state. Since it takes time for each gene product to accumulate or to be degraded, the level of each product in turn rises and falls repeatedly and the system cannot equilibrate. Instead it cycles endlessly through the on and off states of all three genes.

This oscillatory behavior of a mutually repressing circuit was first shown for the so-called repressilator, diagrammed in figure 3.6. The repressilator is a synthetic regulatory circuit consisting of several known transcriptional regulators. When it was engineered into *E. coli* [6], it provided an early demonstration of how a transcriptional circuit might generate temporal oscillations like those of the circadian clocks that control diurnal rhythms in plants and animals. The circuit employs the genes and promoters for three transcriptional repressors, cI (known as the λ-repressor), LacI (the repressor of the *lac* operon), and TetR (the Tet repressor), arranged into a cyclic negative feedback loop. LacI inhibits transcription of *tetR*, whose product TetR inhibits *cI*, which in turn inhibits *lacI* . When cells containing the repressilator system were grown in culture, the activity of the three genes oscillated, with mutual phase shifts and a period of a couple of hours. The oscillations can be tracked optically by placing a GFP reporter under the control of one of the repressors, as shown in figure 3.6.

Stable circadian cycles do appear in individual bacterial cells, as was shown for the cyanobacterium *Synechoccocus elongatus* [7]. But although the repressilator was a useful proof of principle for oscillation in a transcriptional network, the simplest circadian oscillators are not based on transcriptional regulation. The oscillator that drives the *S. elongatus* cycle is generated chemically through repeated cycles of phosphorylation and dephosphorylation within a group of enzymes (KaiABC). The oscillatory behavior of these reactions is so robust that the enzymes can be removed from the cell, purified, and placed in a test tube, and their periodic phosphorylation/dephosphorylation will continue until the mixture exhausts its energy source. Their oscillation inside the cell can be studied using a gene reporter for one of the circadian-driven genes; however, note that GFP cannot be used to monitor the oscillator *in vivo*, as the fluorescence excitation light may itself play the role of a light stimulus and reset the organism's circadian clock. Instead the authors of the *S. elongatus* study used a bioluminescence reporter, so that the cells would emit their own faint light as a readout of the oscillation.

Nonlinear schemes employing feedback or feedforward control, such as those discussed here, are common in bacterial transcriptional regulation. As will be seen later, they enable some very sophisticated regulatory schemes, especially when they also integrate environment cues as well as stochasticity.

References

[1] Pirson I, Fortemaison N, Jacobs C, Dremier S, Dumont J E and Maenhaut C 2000 The visual display of regulatory information and networks *Trends Cell Biol.* **10** 404–8

[2] Willey J, Sherwood L and Woolverton C 2009 *Prescott's Principles of Microbiology* (New York: McGraw-Hill)

[3] Cai L, Friedman N and Xie X S 2006 Stochastic protein expression in individual cells at the single molecule level *Nature* **440** 358–62

[4] Ozbudak E M, Thattai M, Kurtser I, Grossman A D and van Oudenaarden A 2002 Regulation of noise in the expression of a single gene *Nat. Gen.* **31** 69–73

[5] Son M, Ahn S-J, Guo Q, Burne R A and Hagen S J 2012 Microfluidic study of competence regulation in Streptococcus mutans: environmental inputs modulate bimodal and unimodal expression of comX *Mol. Microbiol.* **86** 258–72

[6] Elowitz M B and Leibler S 2000 A synthetic oscillatory network of transcriptional regulators *Nature* **403** 335–8

[7] Mihalcescu I, Hsing W and Leibler S 2004 Resilient circadian oscillator revealed in individual cyanobacteria *Nature* **430** 81–5

IOP Concise Physics

The Physical Microbe
An introduction to noise, control, and communication in the prokaryotic cell
Stephen J Hagen

Chapter 4

Stochastic gene expression

The deterministic models of the previous chapter cannot truly describe protein synthesis at the cellular level because they smooth over the granular character of the underlying molecular processes. If n_A is only the solution to a differential equation such as (3.1) then it can take any value. But if it is the number of copies of a molecule then it must be an integer. Even if the solution to the differential equation for n_A gives a good description of the mean number of copies of A across a population of cells, it cannot capture the discreteness of n_A. In addition, the protein (or RNA) production and degradation processes represented by k and γ in figure 3.4 really are not steady fluxes, but are stochastic molecular processes. When copy numbers are small, as often occurs with transcription factors present at nanomolar concentrations, the resulting noise in $n_A(t)$ and the departure from deterministic model predictions become significant.

4.1 Variability at low copy number

The idea of noise and variability in gene expression had been investigated theoretically [2] and even experimentally [3] in the 1990s and earlier, but its study did not begin in earnest until green fluorescent protein (GFP), followed by other fluorescent proteins, became available as gene reporters, starting in the mid 1990s [4]. Fluorescent protein labeling can be used to measure copy numbers for both mRNA and protein inside individual bacterial cells. The yellow fluorescent protein (YFP), for example, can be detected at the level of individual molecules in living cells. If a *yfp* sequence is fused to a gene, the resulting gene product carries a YFP label that can be detected optically, and the number of copies of that gene product counted. Taniguchi and coworkers [1] generated *yfp* fusions for more than 1000 different genes in *Escherichia coli*: they created a set of strains in which each strain tagged one of its own proteins with a YFP label. Their images, shown in figure 4.1, reveal a striking cell-to-cell variability in protein copy number.

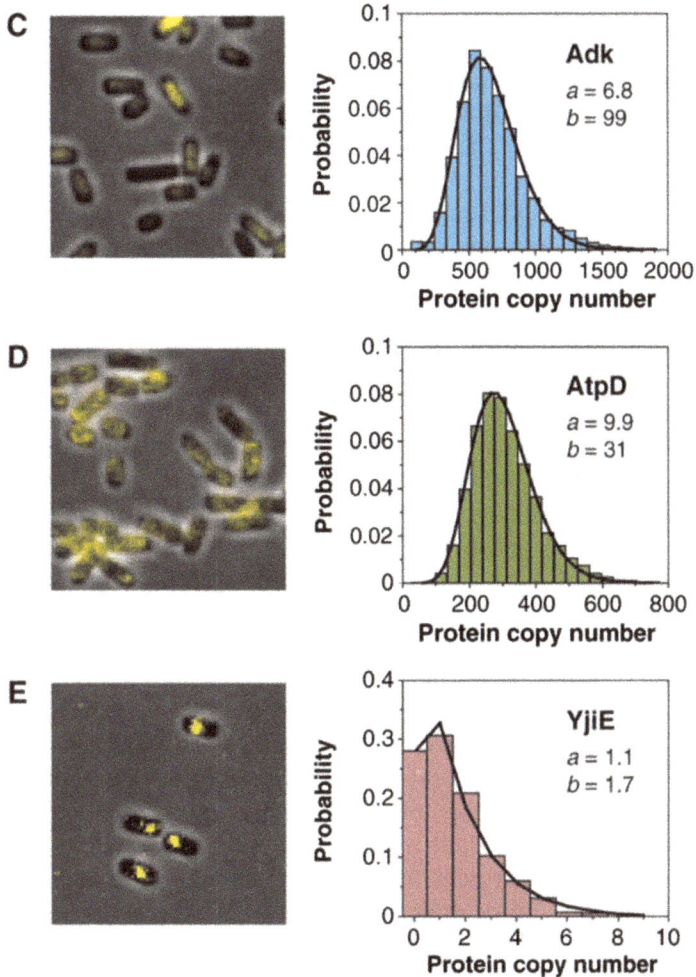

Figure 4.1. The activity of a given gene varies from one cell to another. In the fluorescence images (left), the activity of three genes (*adk*, *atpD*, and *yjiE*) is observed in individual *E. coli*. The cells are engineered to tag the proteins Adk, AtpD, and YjiE with YFP, allowing the proteins to be observed and individually counted in microscopy images of living cells. Adk and AtpD are cytoplasmic and membrane proteins respectively and therefore appear well dispersed in the cells, while YjiE binds to DNA and localizes at the chromosome. Histograms of copy numbers (right) in a population of cells show that the copy number for any one protein is highly variable. Its distribution is also highly variable from one protein to another. The solid curves are fits to the gamma distribution, (equation (4.9)), with fit parameters *a* and *b* as indicated. From [1]. Reprinted with permission from AAAS.

The combination of fluorescent protein reporters and single-cell manipulation technologies like microfluidics has allowed experimentalists to target genes of interest in bacteria and then visualize, using fluorescence microscopy, expression levels in individual cells. Using microfluidics one can isolate cells and study how a behavior controlled by a particular protein varies among cells with different copy numbers of that protein. An example in *E. coli* is the action of the LacZ β-galactosidase, an enzyme that is encoded in the *lac* operon mentioned above. The cell uses the enzyme to break down the sugar lactose. In bulk samples (billions of cells) the activity and abundance of the enzyme is readily measured because β-galactosidase also breaks down certain fluorogenic[1] substrates[2] (that possess a glycosidic bond, the target of the enzyme) to yield a fluorescent product. To study its action in individual cells, Cai *et al* trapped individual *E. coli* within micron-sized (100 pl volume) chambers and supplied them with the fluorogenic dye [5]. Each cell then generated fluorescent products and released them into its environment, causing the fluorescence of its chamber to increase at a rate that depended on the number of copies of β-galactosidase present. That is, the rate of increase of fluorescence in a chamber revealed the number of β-galactosidase molecules possessed by the inhabiting cell. Figure 4.2 shows a histogram of the rates detected in different chambers. Several distinct peaks appear, indicating that different cells contain different numbers of enzyme copies. Further, the histogram is consistent with a Poisson distribution in the copy number, with the average cell having only 0.7 copies. Not only is the average copy number small, but it is less than one!

Looking across many genes, the Taniguchi *et al* data in figure 4.3 show a remarkable variability in copy number. The average copy numbers of different proteins range from 0.1 to 10^4 per cell. Fully half of the studied proteins were present at fewer than 10 copies per cell. For a subset of the same genes, the same authors also applied a fluorescent probe that allowed them to count mRNA copy numbers. These showed that mRNA copy numbers are small, even for highly expressed proteins. The mean abundance of mRNA was 0.05 to 5 copies per cell. Further the mRNA copy number appeared uncorrelated with protein copy numbers. Proteins that were present at very high copy numbers (1000 or greater in figure 4.3) were also not likely to have more abundant mRNA. Although this lack of correlation was unexpected, mRNA and protein act on very different time scales. mRNA is continuously produced and degraded, but protein typically has a longer lifetime, comparable to the cell division time τ_d. The protein copy number at any instant is therefore the accumulated result of many mRNA synthesis and degradation events. The following sections will explore these phenomena in more detail.

4.2 Modeling stochastic expression

Because experimental data show that real proteins are often present at small copy numbers, one should reconsider the deterministic approach of (3.1) and address the size of the statistical fluctuations. Instead of trying to find the average number of copies of A, consider an explicitly statistical viewpoint on production and degradation. Let $P(n, t)$ represent the probability that n copies are present at time t. Then interpret the synthesis rate k as the rate of a Poisson random process. This means that $k\Delta t$ represents the probability that a new copy of A is produced ($n \rightarrow n + 1$) during any short time interval Δt. Similarly interpret the degradation rate γ as

---

[1] A fluorogenic molecule is one that is capable of giving rise to fluorescence.

[2] The substrate is the target molecule whose reaction the enzyme catalyzes to produce a product. Catalase acts on its substrate, H_2O_2, to generate its products, H_2O and O_2.

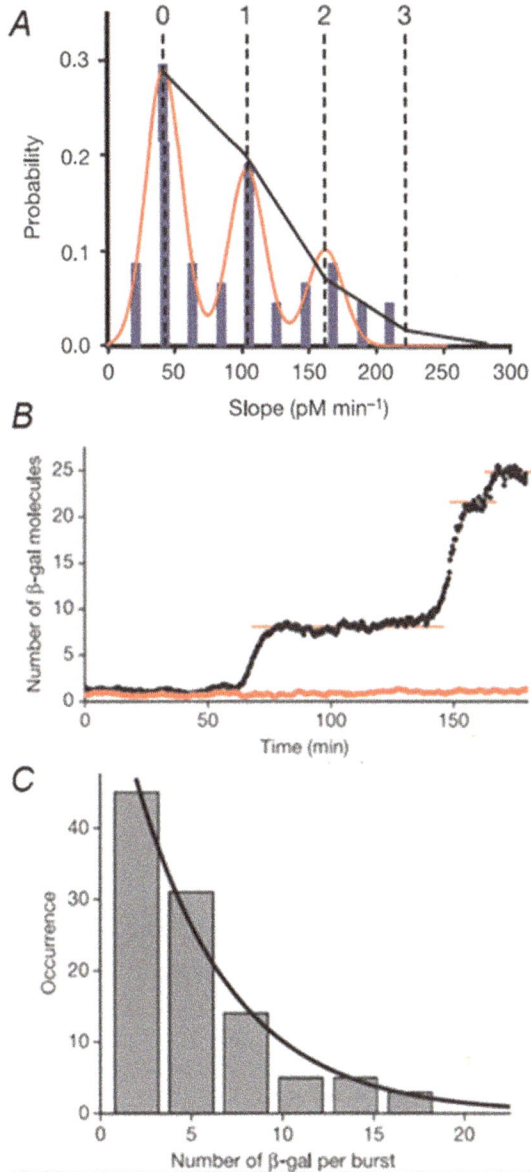

Figure 4.2. (A) Variations in enzyme activity among individual *E. coli*. In [5], each cell occupies a microscopic chamber that retains the products of its enzyme reactions. Cells produce the enzyme β-galactosidase, which breaks down a dye (FDG) to produce a fluorescent product. Enzyme activity is measured by the rate (pM min^{-1}) at which reaction product accumulates in each cell's chamber. The peaks in the histogram correspond to cells that have $n = 0, 1, 2 \ldots$ molecules of the enzyme. The solid black curve is the prediction of a Poisson distribution for the copy number n, with an average $\langle n \rangle = 0.7$. (B) In an individual cell, the number of β-galactosidase copies (black) jumps discontinuously at irregular intervals, as the cell undergoes bursts of protein synthesis. The red curve shows the signal background. (C) A histogram of the number of β-galactosidase copies produced per burst is consistent with an exponential probability distribution (solid curve). Reprinted by permission from Macmillan Publishers Ltd: [5], copyright 2006.

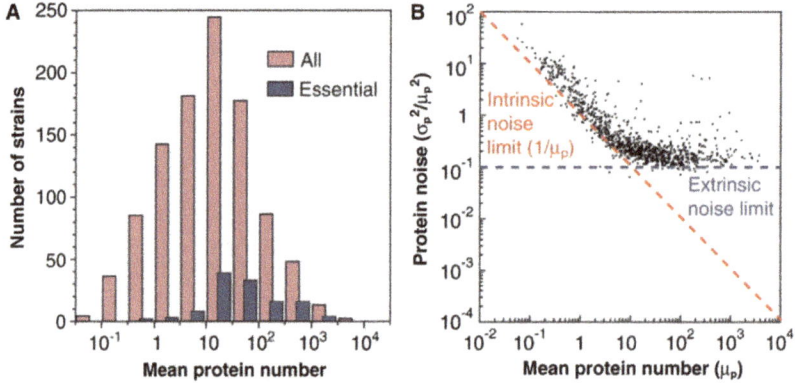

Figure 4.3. (A) Measurement of average copy number, per cell, for 1018 different proteins in *E. coli*. The abundance of each protein was determined from the fluorescence of a strain that was engineered to attach a YFP tag to each copy of the particular protein. Many proteins are present at only 1–10 copies per cell, or even fewer, indicating that the fluctuation in copy number *n* is often similar to (if not greater than) its average value. Essential genes—those required for survival of the organism—are expressed at a higher level than non-essential genes. (B) Cell-to-cell variability in copy number rises rapidly when protein copy number decreases, as expected from a simple Poisson model of intrinsic noise. Extrinsic noise is always present however, and becomes predominant for proteins expressed at high copy numbers. From [1]. Reprinted with permission from AAAS.

describing a Poisson process. The probability is $\gamma \Delta t$ that any given copy of A will degrade ($n \rightarrow n - 1$) during a brief interval Δt.

Then if Δt is a very short time interval, there are three mutually exclusive ways that the cell can have n copies at $t + \Delta t$.

1. There were $n - 1$ copies at time t (which has probability $= P(n - 1, t)$) and a new particle was created during Δt (probability $= k \Delta t$). The joint probability is $P_1 = P(n - 1, t)k\Delta t$.
2. There were previously $n + 1$ copies at time t (probability $= P(n + 1, t)$) and one particle was lost during Δt (probability $= (n + 1)\gamma \Delta t$). The joint probability is $P_2 = P(n + 1, t)(n + 1)\gamma \Delta t$.
3. There were previously n copies (probability $= P(n - 1, t)$) and no production or degradation occurred (probability $= 1 - k\Delta t - n\gamma \Delta t$). The joint probability is $P_3 = P(n - 1, t)(1 - k\Delta t - n\gamma \Delta t)$.

The probability that the cell has n copies at $t + \Delta t$ is then the sum of the probabilities P_1 to P_3,

$$P(n, t + \Delta t) = P(n, t)(1 - k\Delta t - n\gamma \Delta t) \\ + P(n - 1, t)k\Delta t - P(n + 1, t)(n + 1)\gamma \Delta t. \tag{4.1}$$

This difference equation can be rearranged in the limit $\Delta t \rightarrow 0$ to give a differential equation for the evolution of the probability $P(n, t)$.

More simply, however, one can just ask what is the probability distribution for n in a steady state, where $P(n, t)$ does not change over time. This requires $P(n, t + \Delta t) = P(n, t)$ or, based on the above,

$$0 = -P(n, t)(k + n\gamma) + P(n - 1, t)k - P(n + 1, t)(n + 1)\gamma, \qquad (4.2)$$

where Δt has canceled out. If the system is in steady state $P(n, t)$ is replaced by $P(n)$. Defining $\langle n \rangle = k/\gamma$ gives the steady state condition[3]

$$0 = -P(n)(\langle n \rangle + n) + P(n - 1)\langle n \rangle - (n + 1)P(n + 1). \qquad (4.3)$$

One can show that, for this to be true for all n, $P(n)$ must be a Poisson distribution in n with a mean $\langle n \rangle$. It is easier, however, just to plug the Poisson distribution formula,

$$P(n) = \frac{\langle n \rangle^n \exp(-\langle n \rangle)}{n!} \qquad (4.4)$$

into (4.3) and verify that it satisfies the steady state condition.

Therefore the copy number n for an unregulated gene in its steady state turns out to be a Poisson-distributed random variable. The mean copy number $\langle n \rangle = k/\gamma$ is the same as the steady state value found from the deterministic model. Because the variance of the Poisson distribution is equal to its mean[4] this simple model supports the expectation that n will show a great deal of variability, in particular when average expression levels are low.

Exercise (Poisson distribution at steady state) *Verify that the Poisson distribution (4.4) satisfies (4.3).*

Exercise (Continuum limit for $P(n)$) *If $\langle n \rangle$ in (4.3) is very large, n should act like a continuous variable, rather than a discrete one. Then (4.3) should be recast as a differential equation for $dP(n)/dn$, instead of a difference equation involving $P(n + 1)$ and $P(n - 1)$. Find that differential equation.*

4.3 Bursts of gene expression

The model for protein synthesis as a Poisson process can be tested by counting individual copies of protein in a cell. However, protein production involves both transcription and translation steps, not included in the above model, and so one should not too hastily assume that protein copy number will exhibit a Poisson distribution. By contrast mRNA production seems more likely to follow a one-step synthesis model like that of (3.1), and therefore exhibit Poisson statistics. Golding et al [6] studied the actual kinetics of mRNA production in individual E. coli. They used a fluorescent labeling method to tag individual mRNA copies as they were produced, and they counted mRNA copies directly in real time.

[3] Note that k/γ is the steady state copy number in the deterministic model (3.1).
[4] Recall that the Poisson distribution $P(n)$ for a random variable n has several signature properties. One is that its variance $\langle (n - \langle n \rangle)^2 \rangle$ is equal to its mean $\langle n \rangle$. (The brackets $\langle \rangle$ as always denote the average over the distribution.)

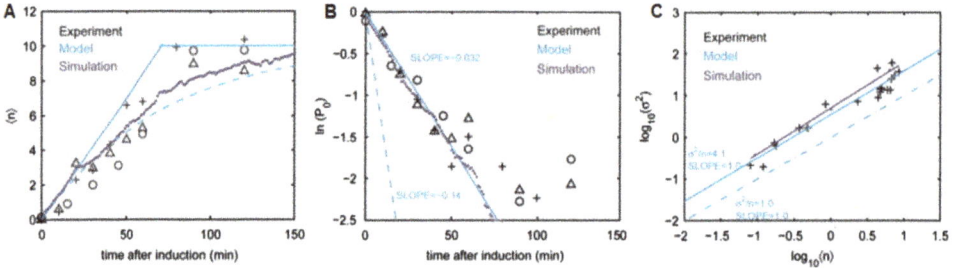

Figure 4.4. Evidence for bursting gene expression, as seen from measurements of mRNA copy number in *E. coli*. (A) The symbols indicate the average number $\langle n \rangle$ of mRNA transcripts per cell, following induction of a gene at $t = 0$, over several independent experiments. The steady state mRNA copy number appears to be roughly 10, and the copy number approaches the steady state with an exponential relaxation time constant of roughly 70 min, leading to an estimated synthesis rate $k \simeq 0.14$/min and a degradation rate $\gamma \simeq 0.014$/min, by (4.5). The dotted curve is the result of a simulation of (3.1), using these k and γ values. (B) In the exponential model of (4.6), the proportion of cells containing no mRNA transcripts at t should decrease as $\exp(-kt)$. Although the proportion does decrease exponentially, it decreases at a rate $\simeq 0.032$/min, four times slower than k. The solid curve shows a simulation of the bursting model. (C) As gene induction is varied (by addition of the inducer IPTG) the variance in the mRNA copy number is proportional to, but greater than the mean copy number. The proportionality constant is approximately four, consistent with a model in which mRNA transcripts are generated in bursts of four. Reprinted from [6]. Copyright 2005 with permission from Elsevier.

In the Golding *et al* experiment, production of the tagged mRNA was placed under control of the *lac* repressor, an inducible repressible system, so that its transcription could be switched on at a particular time by addition of an inducer[5]. Then, applying the deterministic model to the production of mRNA (where k is the rate of transcription and γ is the rate of mRNA degradation), the population-averaged mRNA number should grow as

$$\langle n \rangle = \frac{k}{\gamma}(1 - e^{-\gamma t}), \tag{4.5}$$

as in (3.2), where t is the time since induction. Figure 4.4(A) shows that $\langle n \rangle$ does grow in this way, relaxing toward its steady state in exponential fashion. A fit to the data gives the values of k and γ.

Knowing those parameters, one can then perform a stochastic simulation[6] of (4.5) and see how much variability in $n(t)$ is expected. This is akin to finding a numerical solution for $P(n, t)$. (The figure shows the data, the deterministic result, and simulation.)

One property of a stochastic model that does not arise in the deterministic model is that some cells by chance will be slow to produce mRNA. There is a probability p

---

[5] Recall that the *lac* operon is an inducible repressible system, figure 3.3. A gene can be placed under control of LacI, the *lac* repressor, by replacing its promoter with P*lac*, the promoter of the *lac* operon. If the lac repressor is present, transcription of the gene will be repressed unless an inducer is provided. In the laboratory the allolactose analog IPTG (isopropylthio-β-D-galactoside) is commonly used to induce *lac*-controlled genes.
[6] The classic method for simulating a stochastic chemical reaction is that of Gillespie [7].

that, at a time t after induction, a given cell will not yet have made any mRNA. This probability should decrease exponentially according to

$$p(t) = e^{-kt}. \tag{4.6}$$

As k is known, the fraction of cells lacking mRNA in the experiment should match (4.6). However, although figure 4.4(B) does show an exponential decay of p, the number of cells lacking any mRNA transcripts is consistently higher than predicted. p decreases at a rate only $\sim 1/4$ as great as predicted by the k extracted from the mRNA production data, figure 4.4(A). The exponential behavior of p agrees with the simple transcription model, but its actual rate does not.

Because the mRNA copy number should follow a Poisson distribution, the authors also tested whether the variance σ_n^2 in the mRNA copy number n was equal to the mean, $\langle n \rangle$. Again the functional behavior is correct but the numerical value is not: as figure 4.4(C) shows, σ_n^2 is consistently proportional to $\langle n \rangle$ as the induction level is varied, but the two quantities are not equal. Instead the variance is about 4× larger than the mean.

In short, the mRNA copy number shows some properties expected for the Poisson production model, but the numbers seem to miss by a factor of about four.

A straightforward explanation for these discrepancies is simply that the Poisson process that is being observed is not the transcription of individual mRNAs, but rather the transcription of bunches of mRNAs. Rather than occasionally producing a single transcript, the gene occasionally produces a burst of transcripts. Then it is the number of bursts that have occurred, not the number of transcripts, that should be Poisson distributed. If there are m transcripts per burst, then the number n of transcripts in a cell is $n = mq$ where q is the number of bursts present. Because of the Poisson behavior, the variance in q is $\sigma_q^2 = \langle q \rangle$. The mean and variance of q are related to those of n by $\langle n \rangle = m\langle q \rangle$ and $\sigma_n^2 = m^2 \sigma_q^2$. Therefore $\sigma_n^2 = m\langle n \rangle$. From this viewpoint the data are consistent with a bursting model; transcription is a Poisson process, but each transcription event generates about $m = 4$ transcripts.

The β-galactosidase data of figure 4.2 tell a similar story. They not only show that copy number varies from one cell to another, but they also show the irregular kinetics of protein synthesis itself. The protein is produced in bursts of 10–20 molecules at a time. Figure 4.2(C) shows that the size of these bursts—the number of protein copies produced—is exponentially distributed.

Transcriptional bursting is an important source of noise in gene expression, not only in bacteria but also in eukaryotes. In yeast, for example, genes switch randomly between transcriptionally active and inactive states. In higher eukaryotes transcriptional bursting can result from the structure and dynamics of chromatin, which by opening or closing can physically control whether genes are accessible for transcription. Transcriptional bursting is important even in mammalian cells, where we might expect very sophisticated regulation of mRNA copy numbers [8]. In bacteria the mechanisms of transcriptional bursting are still under investigation, although it is very plausibly a consequence of the buildup of physical, supercoiling tension in the DNA molecule during transcription, leading to a shutdown of transcription initiation [9]. Enzymes known as gyrases that cut the DNA and release its accumulated strain can allow transcription to resume.

Finally, it is fairly easy to understand the exponential histogram of burst sizes in figure 4.2(C) if each burst corresponds to multiple rounds of translation from a single mRNA copy. Consider that each individual mRNA will be available for translation until it is destroyed by ribonuclease E (RNaseE), the enzyme that breaks down most mRNA in *E. coli*. Let $p(t)$ be the probability that an mRNA present at $t = 0$ still exists at time t. Because degradation occurs at a rate γ_R (figure 3.4), the probability that the transcript is *not* destroyed during an interval dt is $1 - \gamma_R \, dt$. Then

$$p(t + dt) = p(t)(1 - \gamma_R \, dt). \tag{4.7}$$

As $dt \to 0$ this leads to

$$\frac{dp}{dt} = -\gamma_R p(t). \tag{4.8}$$

Integrating this expression[7] gives $p(t) = \exp(-\gamma_R t)$. Then the probability that the mRNA survives for exactly t but is degraded during the following brief dt is $\exp(-\gamma_R t)\gamma_R \, dt$. As a result the lifetimes of the mRNAs are exponentially distributed, with an average lifetime $1/\gamma_R$. As long as translation from each mRNA occurs at a reasonably brisk rate k_P (protein copies/min/mRNA) relative to the mRNA lifetime, the number of protein copies generated from each mRNA will also be exponentially distributed, and the average protein burst will consist of k_P/γ_R copies[8].

Exercise (mRNA production) *Use an argument similar to that preceding (4.1) to show that the probability $p(t)$ that a cell has not yet made any mRNA must obey (4.6). Write an expression for $p(t + dt)$ in terms of $p(t)$, k and dt where dt is a very short time interval. Show that it leads to a differential equation for dp/dt whose solution is (4.6).*

4.4 Protein distributions with both transcription and translation

The fact that each mRNA can produce a burst of protein copies, while the mRNA is produced through its own random process, underscores the earlier point that protein production is not likely to obey the model that predicted a Poisson distribution. One should anticipate that fluctuations in protein number may be larger than the Poisson model predicts.

Therefore it is worth revisiting the two-state deterministic model discussed earlier, (3.4) and (3.6), in which transcription and translation are treated as distinct Poisson processes with separate rates. The same average, steady state protein copy number

---

[7] Note the initial condition $p(0) = 1$.
[8] Perhaps it is more precise to say that the number of protein copies per mRNA follows a geometric distribution, rather than an exponential distribution. Protein copy number is an integer and so the analysis should strictly use a discrete (geometric) distribution rather than a continuous (exponential) distribution.

n_A can be achieved by efficient translation and inefficient transcription, or by the reverse. Ozbudak and co-workers [10] showed that stochasticity makes a difference here, because while the average n_A may be the same in both cases, the variability, whether in one cell over time or from cell to cell, can be very different.

To demonstrate the different noise consequences of transcription and translation, they modulated both the transcription rate k_R and the translation rate k_P of a *gfp* reporter in *Bacillus subtilis*. They controlled the rate of transcription by placing the *gfp* reporter under control of an inducible promoter. They separately manipulated the efficiency of translation by inserting mutations into the ribosomal binding site on the promoter for the *gfp*. They found that transcription and translation have very different effects on the cell-to-cell variability in GFP production. When transcription is inefficient but translation is efficient, the number of mRNA copies tends to be small, with large statistical fluctuations. These fluctuations are amplified by the efficient translation process. Then fluctuations in protein copy number are large. By contrast, when transcription is efficient but translation is inefficient, there is a large pool of mRNA transcripts available for translation, subject to less fluctuation, where each of these produces few protein copies. In this case fluctuations in protein copy number tend to be smaller. Transcription makes a larger contribution to noise in gene expression than does translation.

So what is the expected probability distribution of protein copy numbers in the two-step (transcription + translation) model for stochastic gene expression? If mRNA transcripts are produced by a Poisson process and the number of copies of protein produced per mRNA follows an exponential distribution as discussed above, then the number of protein copies present in the cell is the sum of several random numbers drawn from an exponential distribution. More precisely, it is the sum of all the protein copies produced by all the mRNAs produced over the longest relevant time scale. The long time scale is the lifetime of one protein, $1/\gamma_P$[9]. There will be on average a transcripts present at one time, where $a = k_R/\gamma_P$ is the number of mRNA transcripts produced during the lifetime ($1/\gamma_P$) of a protein copy. The average number of protein copies produced during an mRNA transcript's lifetime is then $b = k_P/\gamma_R$. The actual number of protein copies produced per transcript is exponentially distributed, so the number n of protein copies in the cell will be the sum of a values of an exponentially distributed variable that has average value b. The probability distribution for such an n is a gamma distribution[10]

$$P(n) = \frac{n^{a-1} \exp(-n/b)}{\Gamma(a)b^a}, \tag{4.9}$$

where Γ is the gamma function [5]. The parameters a and b, which are defined by the transcription and translation parameters above, relate to the mean μ and variance σ^2

---

[9] If mRNA were longer-lived, then the relevant time scale would be $1/\gamma_R$.

[10] The literature sometimes gives $P(n)$ as a negative binomial distribution instead of a gamma distribution. The negative binomial is the discrete version, arising from a geometrically distributed number of protein copies per transcript. The gamma is the continuous version, arising from an exponentially distributed number of protein copies per transcript [5].

of the distribution by $\mu = ab$ and $\sigma^2 = ab^2$. The mean protein copy number is then $\mu = ab = k_p k_R / \gamma_P \gamma_R$ (as it is in the deterministic version of the two-step model, equation (3.6)).

Note that the gamma distribution has two parameters, unlike the Poisson distribution. While a has a large effect on the overall shape of $P(n)$, b acts more to set the scale for n. This should make sense given the above interpretation of these parameters. As with the Poisson model, more active genes should be relatively less noisy, as the ratio of the variance to the squared mean decreases at high transcription rates, $\sigma^2/\mu^2 = 1/a \propto 1/k_R$. As the *E. coli* examples in figure 4.1 show, the gamma distribution generally provides a rather good fit to real protein copy number distributions.

4.5 Intrinsic and extrinsic noise

Noise in gene expression is a well established fact. As to its origin, we have treated it so far as a statistical effect. But in the Taniguchi *et al* data of figure 4.3, the gene expression noise σ, measured as the standard deviation of the protein copy number[11], is qualitatively different for low expression (mean copy number $\mu < 10$) and high expression ($\mu > 10$) genes. At low expression levels the ratio of variance to squared mean, σ^2/μ^2 varies inversely with μ. That is, the protein copy number distribution generally agrees with the Poisson or gamma distribution. However, as expression levels increase, σ^2/μ^2 does not continue to decrease. Instead it flattens out above $\sigma^2 \simeq 0.1\ \mu^2$, indicating that even the most robustly expressed proteins are still subject to a noise level of at least $\sigma \simeq 0.3\ \mu$, or roughly 30%. This implies a changing mechanism of noise; there exist different sources of variability in gene expression, and the dominant sources for one gene may differ from those for another.

One source of variation is the noise that is intrinsic to the statistical nature of production and degradation, as discussed so far. This intrinsic noise must affect the copy number of any protein, even if every other property of the cell is (ideally) predictable.

But another form of noise is so-called extrinsic noise, originating from variability in all the other, external influences on transcription and translation. The number of ribosomes, the number of plasmid copies present, the concentrations of regulatory molecules and amino acids, and various environmental cues can all affect the expression of a gene, and they will affect one cell differently from another. They add an extra layer of variability to gene expression.

Perhaps it is not apparent that one can identify noise as either intrinsic or extrinsic. Elowitz and co-workers [11] proposed a practical distinction as follows: if a cell carries two copies of the same gene under the control of the same promoter, then the intrinsic noise in these genes is that part of their variation that is statistically uncorrelated, while the extrinsic noise is the part that is correlated. They illustrated this by designing an *E. coli* strain that expressed both the cyan fluorescent protein

---

[11] Noise in expression is also sometimes quantified by the coefficient of variation, 'CV', defined as $CV = \sigma/\mu$. Again σ is the standard deviation and μ is the mean. Note that if the copy number follows a Poisson distribution then $CV = \mu^{-1/2}$.

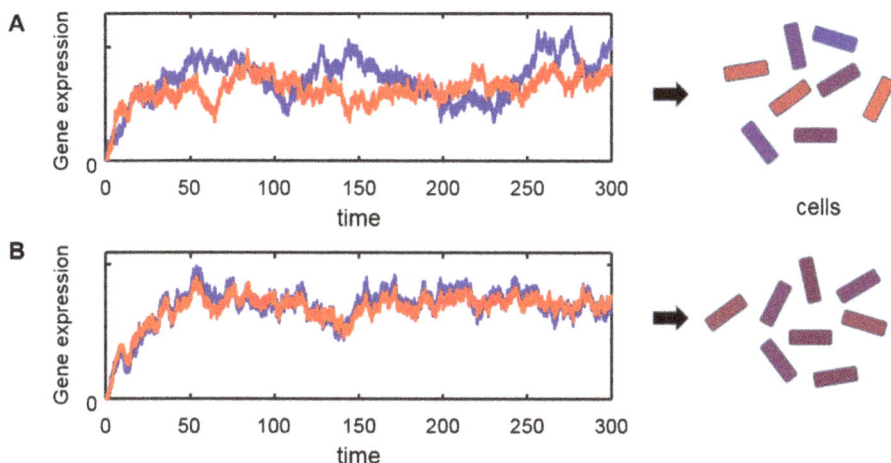

Figure 4.5. The distinction between intrinsic and extrinsic noise in gene expression is illustrated by a hypothetical experiment. A cell is induced at time $t = 0$ to express two genes that produce red and blue reporters, respectively, at equal average rates. (A) If the noise in expression is primarily intrinsic, then blue and red levels fluctuate independently. Individual cells are highly variable in color as a result. (B) If the noise is primarily extrinsic, then blue and red protein levels fluctuate in correlated fashion, leading to more uniform color in the population of cells. Note that the figure shows two different manifestations of noise: there is cell-to-cell variation in the population (right) as well as temporal variation (left) within each cell. The figure is based on [11].

(CFP) and YFP from identical promoters located at different sites on the chromosome. The relative activity of the two copies could then be assessed from the color of the cell's fluorescence. If intrinsic noise is small, extrinsic noise should affect both genes synchronously, causing fluctuations in total protein over time, and from one cell to another. Yet the relative proportions of CFP and YFP should remain mostly constant, leading to a generally uniform cell color. This scenario is illustrated schematically in figure 4.5(A). If, however, intrinsic noise greatly exceeds extrinsic noise, then CFP and YFP should vary in uncorrelated fashion, leading to color variation among different cells and in a given cell over time. The idea is illustrated in figure 4.5(B). This behavior was in fact observed experimentally. Tuning the strength of noise, Elowitz *et al* were able to generate uncorrelated (intrinsic) or correlated (extrinsic) fluctuations in the CFP and YPF protein levels.

This two-reporter scheme also allows the relative contributions of extrinsic and intrinsic noise to be quantified over a range of expression levels. As shown in the scatter plot in figure 4.6(A), the cell-to-cell variation in expression of two independent reporters can be divided into correlated (extrinsic) and uncorrelated (intrinsic) components. The intrinsic and extrinsic parts can be extracted quantitatively by measuring variance in expression of the two reporters as follows:

- The total noise η_{total}^2 is the total, cell-to-cell variance measured in the expression of both reporters.
- The intrinsic noise η_{int}^2 is defined as the mean squared distance of the data in the scatterplot from the diagonal line (representing equal expression of both reporters), in figure 4.6A.

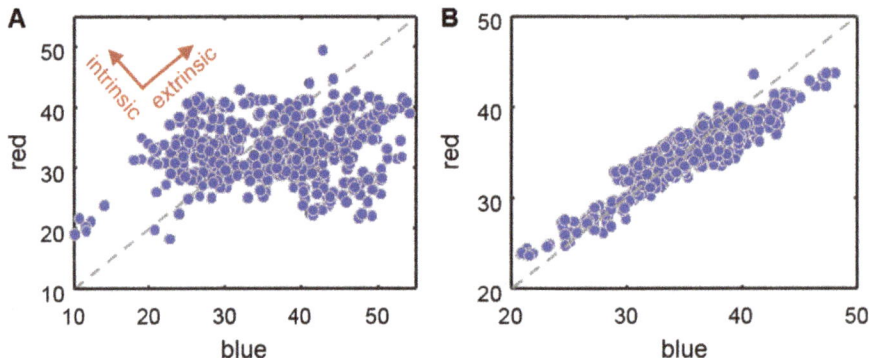

Figure 4.6. In this simulation based on the hypothetical experiment of figure 4.5, scatter plots compare the expression of two reporters, denoted red and blue, in a population. Each point shows the expression levels in one cell. (A) When intrinsic noise is large, its uncorrelated character leads to a broad scatter from the diagonal (red = blue) line. Intrinsic noise is defined as the mean squared distance of the points from the diagonal. (B) When extrinsic noise is larger than intrinsic noise, cells are scattered along the diagonal but remain close to it. The figure is based on [11].

- The extrinsic noise η_{extr}^2 is simply that part of the total noise that is not intrinsic. It is obtained from the other quantities through the expression

$$\eta_{\mathrm{total}}^2 = \eta_{\mathrm{int}}^2 + \eta_{\mathrm{extr}}^2. \tag{4.10}$$

Independent parts of the noise add as squares[12], as in the Pythagorean theorem.

If the noise contributions in (4.10) are obtained [11] experimentally for a particular reporter system under a particular condition, their relative magnitudes can then be determined as a function of overall gene expression activity. Elowitz *et al* used a *lac*-repressor system in which they could dial up or down the expression of the CFP and YFP reporters and see how η_{int} and η_{extr} respond. They found as expected that η_{int} varies inversely with expression level, but η_{extr} does not fall monotonically as expression increases. In fact it may initially rise as gene activity increases, before falling at higher expression levels. More generally the data of Taniguchi *et al* indicate that extrinsic noise remains significant for all genes at high expression levels.

4.6 Noise reduction and stability through feedback

Increasing protein production is one way a cell can reduce the noise in expression, relative to the average. Another is to regulate expression using negative feedback. Suppose that a gene produces a protein whose concentration is R. If the gene is completely unregulated then its rate of expression depends on the interaction of the RNA polymerase with the promoter. The affinity of the polymerase can be

---

[12] This is also known as addition in quadrature.

characterized by a parameter k_P and the concentration of polymerase can be denoted P, so that at any instant there is a probability

$$\frac{k_P P}{1 + k_P P} \tag{4.11}$$

that the gene is being transcribed. Then in a simple deterministic model [12], where γ is the degradation rate of the protein, the function $f_{nr}(R)$ describes the rate of change of R,

$$\frac{dR}{dt} = f_{nr}(R) = a\frac{k_P P}{1 + k_P P} - \gamma R. \tag{4.12}$$

Here a is a parameter that accounts for the rates of both transcription and translation, relating the rate of protein synthesis to the interaction status of the polymerase. At a concentration R_0 the system has a steady state $dR/dt = 0$, characterized by $f_{nr}(R_0) = 0$. How does the system respond to noise, which causes R to deviate slightly from R_0? If $R = R_0 + \delta R$, where δR is small, then

$$\frac{dR}{dt} = f_{nr}(R_0 + \delta R) \approx f_{nr}(R_0) + f'_{nr}\delta R \tag{4.13}$$

by Taylor expansion. Because $f_{nr}(R_0) = 0$, $dR = d\delta R$, and $f'_{nr} = -\gamma$, one can rewrite this as

$$\frac{d\delta R}{dt} \approx -\gamma\,\delta R \tag{4.14}$$

for small δR. The system tends to relax toward its steady state at R_0.

Now suppose that the system uses negative autoregulation, so that the gene product R represses its own transcription. This can be modeled by modifying (4.12) to

$$\frac{d\delta R}{dt} = f_{ar}(R) = a\frac{k_P P}{1 + k_P P + k_R R} - \gamma R, \tag{4.15}$$

where k_R characterizes the affinity of the repressor for the operator. If due to noise the protein level departs from its steady state value, $R = R_0 + \delta R$, R will evolve according to

$$\frac{dR}{dt} = f_{ar}(R_0 + \delta R) \approx f_{ar}(R_0) + f'_{ar}\delta R = f'_{ar}\delta R \tag{4.16}$$

by Taylor expansion. f'_{ar} is a little more complex than before,

$$f'_{ar} = \frac{-ak_P k_R P}{(1 + k_P P + k_R R)^2} - \gamma. \tag{4.17}$$

In any case, however, $d\delta R/dt = f'_{ar}\delta R$ will still be negative whenever δR is positive, and so R will tend to relax toward its steady state value. The relative stability of the autoregulated (4.15) and unregulated (4.12) systems, or their relative tendency to return to their steady state, is summarized by the ratio

$$S_r = f'_{ar}/f'_{nr}. \tag{4.18}$$

Becskei and Serrano [12] evaluated this ratio for various parameter values, as sketched briefly in figure 4.7. For virtually all reasonable choices of parameter values, S_r is greater than one, implying that the negative-feedback-regulated gene returns faster to its steady state output R_0 than does an unregulated gene.

The authors tested this simple picture by constructing a self-regulating model system in *E. coli*. The gene for a synthetic protein TetR-GFP (the Tet repressor, fused to GFP) was placed under control of the operator that is repressed by the TetR portion of the protein. The authors then measured cell-to-cell variability in the expression of the gene. Figure 4.7 shows that the negative feedback from the repressor led to a sharp reduction in gene expression levels, but it also reduced

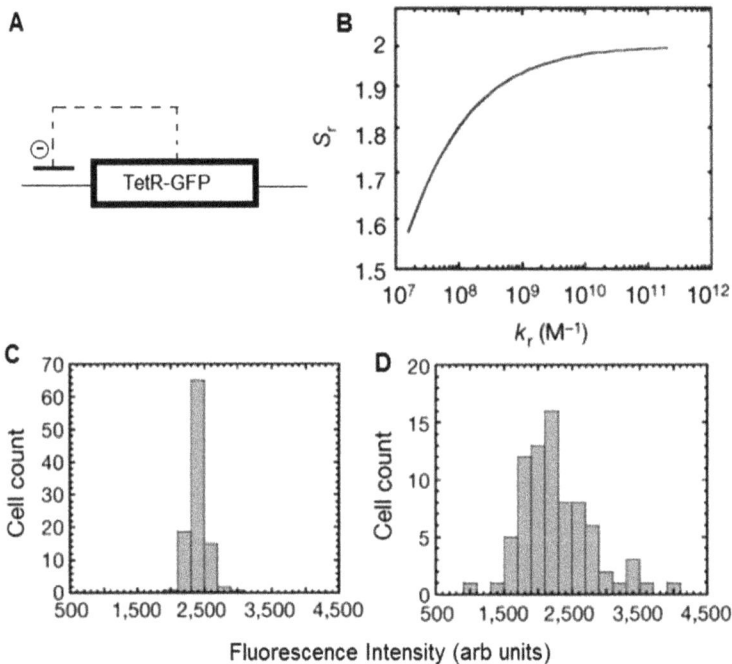

Figure 4.7. The effect of negative autofeedback on gene noise was tested by designing a synthetic gene system that repressed its own transcription. (A) The gene for GFP was fused to the gene for the tetracycline repressor TetR and placed under control of the tetracycline operator. The product (TetR-GFP) binds to the operator, repressing its own transcription while providing a fluorescent readout (via GFP) of its expression level. (B) Modeling of the relative stability parameter S_r, equation (4.18), indicates that an autoregulated (4.15) transcriptional process virtually always has greater output stability than a non-regulated (4.12) process. This is indicated by the fact that S_r is always greater than one. (C) In the experiment where *E. coli* produces GFP that is under negative autofeedback control from a Tet repressor, the coefficient of variation in GFP production is small as long as the negative feedback circuit functions. (D) When negative feedback is weakened by a mutation in the repressor, limiting its affinity for the operator, the noise level, indicated by the width of the histogram, increases. Reprinted by permission from Macmillan Publishers Ltd: [12], copyright 2000.

the variance in gene expression. This is apparent as a reduction in the coefficient of variation, $CV = \sigma/\mu$, in the cell fluorescence histograms.

The effect of negative feedback on noise can be further tested by progressively reducing the affinity of the repressor for its target. Becskei and Serrano did this first by introducing a mutation into the repressor, and then by completely replacing the operator. This led to an increase both in expression and in noise. As figure 4.7 shows, feedback control is highly effective in reducing noise.

However, even though negative feedback is a common design feature in regulatory circuits, noisy gene expression is still ubiquitous. It occurs, as the Taniguchi *et al* [1] data show, in regulated and unregulated genes alike. More to the point, noise probably brings some useful benefits to the organism, such as bringing a certain amount of non-heritable diversity to a population. Although variable gene expression in a population of bacteria may seem inconsistent with optimal growth, it allows a genetically identical population to express a diversity of phenotypes, of which at least a few may enjoy some survival advantage if the environment changes unexpectedly. This idea of 'bet hedging' through noise has interesting applications in the phenomena of persistence, switching, and communication, discussed below.

References

[1] Taniguchi Y, Choi P J, Li G-W, Chen H, Babu M, Hearn J, Emili A and Xie X S 2010 Quantifying *E. coli* proteome and transcriptome with single-molecule sensitivity in single cells *Science* **329** 533–8

[2] McAdams H H and Arkin A 1997 Stochastic mechanisms in gene expression *Proc. Natl. Acad. Sci.* **94** 814–9

[3] Spudich J L and Koshland D E 1976 Non-genetic individuality: chance in the single cell *Nature* **262** 467–71

[4] Chalfie M, Tu Y, Euskirchen G, Ward W W and Prasher D C 1994 Green fluorescent protein as a marker for gene expression *Science* **263** 802–5

[5] Cai L, Friedman N and Xie X S 2006 Stochastic protein expression in individual cells at the single molecule level *Nature* **440** 358–62

[6] Golding I, Paulsson J, Zawilski S M and Cox E C 2005 Real-time kinetics of gene activity in individual bacteria *Cell* **123** 1025–36

[7] Gillespie D T 1977 Exact stochastic simulation of coupled chemical reactions *J. Phys. Chem.* **81** 2340–61

[8] Raj A, Peskin C S, Tranchina D, Vargas D Y and Tyagi S 2006 Stochastic mRNA synthesis in mammalian cells *PLOS Biol.* **4** e309

[9] Chong S, Chen C, Ge H and Xie X S 2014 Mechanism of transcriptional bursting in bacteria *Cell* **158** 314–26

[10] Ozbudak E M, Thattai M, Kurtser I, Grossman A D and van Oudenaarden A 2002 Regulation of noise in the expression of a single gene *Nat. Gen.* **31** 69–73

[11] Elowitz M B, Levine A J, Siggia E D and Swain P S 2002 Stochastic gene expression in a single cell *Science* **297** 1183–6

[12] Becskei A and Serrano L 2000 Engineering stability in gene networks by autoregulation *Nature* **405** 590–3

Chapter 5

Phenotypic switching

Antibiotics are a diverse group of compounds that suppress bacterial populations, either by stalling growth or by killing growing cells outright. Penicillin, for example, is a so-called β-lactam antibiotic. It kills actively dividing bacteria by inhibiting correct synthesis of the peptidoglycan layer of the cell wall. The discovery of penicillin and other antibiotics revolutionized the treatment of bacterial infections and has saved millions of lives. However, antibiotics sometimes fail to kill the bacteria that they target. One reason for failure is that bacterial strains become resistant; they acquire mutations that modify the molecular target of the antibiotic or allow the cell to neutralize the antibiotic or block its import into the cell [1]. Antibiotic resistance has two characteristics: it is a heritable trait, and it increases the 'minimum inhibitory concentration', or the antibiotic concentration needed to suppress growth. Another reason that antibiotic may fail to kill cells is simply that the cells have stopped growing, perhaps because of environmental stress such as starvation, as occurs in stationary phase. In this case the cells may be called tolerant of an antibiotic, although not resistant [2].

There is another mechanism for failure, discovered as early as 1944 by Joseph Bigger [3]. Bigger found that penicillin treatment did not completely sterilize an actively growing culture of *Staphylococcus pyogenes*. Instead, a tiny minority of the *S. pyogenes* cells, perhaps 1 in 10^6, survived their contact with penicillin because they were dormant and not actively dividing, even though growth conditions were otherwise favorable. Such cells are called persisters. Persisters survive because they have ceased growth under conditions where other cells in the same population grow. But if removed from the antibiotic medium and placed into more hospitable medium, they eventually resume normal growth. Therefore their survival trait is not heritable. Most of their progeny grow normally, with only a tiny fraction being persisters.

Persistence is a phenotypic trait, arising from the interaction between the organism's genetic program and its environment. It is not antibiotic tolerance

doi:10.1088/978-1-6817-4529-9ch5

because it is limited to a subpopulation of cells. It is a transient state, because cells can switch into and out of the persister phenotype [4]. That is, it is a form of phenotypic switching, or heterogeneity that arises stochastically among genetically identical cells. Although other types of phenotypic variation occur in bacterial cultures[1], persistence is of special interest because the effectiveness of antibiotics is so important in human medicine and elsewhere. One should then ask what regulatory mechanisms drive persistence, what the dynamics of the phenotypic switch are and how to model them physically. Single-cell observations have contributed a great deal of insight to such studies because persistence is almost by definition rare in the population. It is a case where a tiny minority of cells can play a disproportionate role in the survival of a larger group.

5.1 Two types of persisters

Although persistence is not heritable, certain mutations increase the number of persisters in a culture. In the *hipA7* allele mutant[2] of *Escherichia coli*, persisters are about 1000-fold more common than in the wild type [5]. Figure 5.1 shows an experiment where antibiotic is supplied to a culture containing such a strain. After the antibiotic is added at time $t = 0$ the number of viable cells (measured as CFUs[3]) is measured as a function of time. The viable cells initially decline linearly on this scale, consistent with exponential killing of susceptible cells.

However the killing curve has a biphasic character. After the number of live cells falls very low, the curve flattens, indicating that the most susceptible cells have been killed. The remaining cells—the persisters—die at a far slower rate. If the persisters are removed from the antibiotic, regrown in fresh medium and then treated again with antibiotic, their progeny will generate another similar biphasic killing curve.

The linear portions of the killing curve reflect two different exponential processes of cell death, where (1) normal cells die at a rate k_n (probability of dying, per unit time) and (2) persister cells die at a rate k_p. One may imagine two possible interpretations of the persister death rate k_p. It could reflect the actual rate at which antibiotic kills persisters. Alternatively, it could reflect the rate at which persister cells emerge from the persistent state, resuming normal growth and thus becoming vulnerable to antibiotic. Because experimental evidence suggests that cells emerge very slowly from persistence, it is likely that the observed k_p reflects the rate at which the antibiotic kills persisters [7].

In fact there exist qualitatively different types of persistence, which can be observed in different high-persistence mutants of *E. coli*. One type appears as a response to stress, particularly starvation. So-called type I persisters are triggered to enter dormancy by internal signals that occur during stationary phase. When placed into fresh growth medium, they enter the dormant state and exhibit a pronounced

---

[1] Genetic competence is another example, to be discussed later.
[2] Alleles are alternative or mutant forms of a particular gene, found in different individuals of the same species at the same site on the chromosome. *hipA7* refers to a particular allele of the *hipA* gene in which two important mutations are present. The name of the gene comes from 'high persistence'.
[3] CFU = a colony forming unit, a microbial cell capable of dividing.

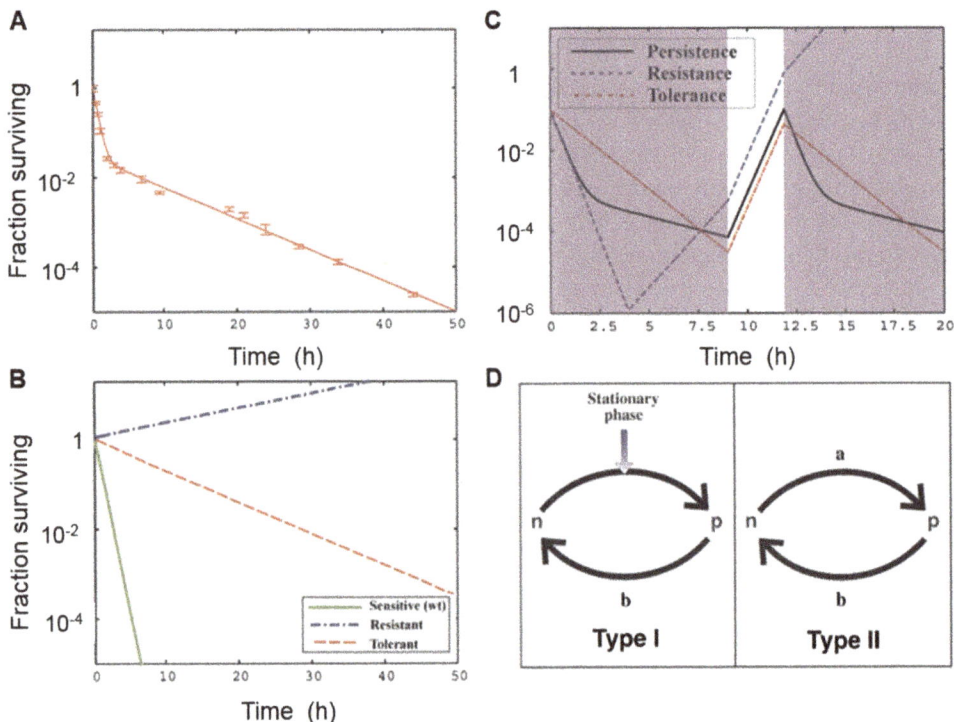

Figure 5.1. The persister phenotype in bacteria, from Gefen and Balaban [6]. (A) When a culture of the *hipA7* mutant strain of *E. coli* is exposed to the antibiotic ampicillin, the number of viable cells declines exponentially over a period of hours until about 1% remain. Subsequently the slope of the curve decreases, indicating a slower death rate among the remaining cells, called persisters. (B) Schematic illustrating the difference between resistant and tolerant strains growing in the presence of antibiotic. The resistant strain grows in the presence of antibiotic. Tolerant cells are killed, but at a lower rate than the wild type (sensitive) strain. (C) Unlike resistance and tolerance, persistence is not a heritable trait. If after a period of antibiotic exposure (first purple region) the antibiotic is removed (white region), the persister cells (black curve) will regrow. When exposed again to antibiotic (second purple region) the persisters will show the same biphasic decline that they showed under the first antibiotic treatment. The tolerant strain shows a population-wide, reduced sensitivity to the antibiotic (red dashed curve). True antibiotic resistance (blue dashed line) emerges as a heritable trait (here at about 3 h) that allows the resistant cells to grow during the second antibiotic treatment. (D) A model for type I and type II persistence, as captured by (5.1). Type I persisters are driven into the persistent state *p* at stationary phase, and emerge slowly to the normal growth state *n*. Type II persisters switch stochastically between *p* and *n* under all conditions, at rates *a* and *b*. Reproduced from [6] by permission of Oxford University Press.

delay or lag before resuming normal growth. The *hipA7* mutation in *E. coli* increases the number of type I persisters to about 1%–2% of the population [6].

A second type of persistence, called type II, is seen when cells that are already growing normally switch spontaneously into a dormant state. The *hipQ* mutation in *E. coli* enhances the fraction of cells showing type II persistence. Thus type I persisters survive antibiotic because they are slow to emerge from stationary phase, while type II persisters survive because they have switched temporarily into a slower growth state [2]. Presumably a wild-type population contains both types of persisters, albeit in very small numbers, leading to a generally complex response to antibiotics.

In the type II scenario where two phenotypes denoted n (normal) and p (persistent) grow at rates λ_n and λ_p, respectively, we can model the time dependence of the populations by assuming a constant rate a of switching from n to p, and a rate b of switching from p to n [6]. Then in the absence of antibiotic

$$\frac{dn}{dt} = -an + bp + \lambda_n n$$
$$\frac{dp}{dt} = +an - bp + \lambda_p p. \tag{5.1}$$

In the presence of antibiotic the cells do not grow, but die at rates k_n and k_p, giving

$$\frac{dn}{dt} = -an + bp - k_n n$$
$$\frac{dp}{dt} = +an - bp - k_p p, \tag{5.2}$$

if the a and b rates are unchanged. For type I persistence as in the $hipA7$ strain, the persistent state is induced by the stationary phase, and so $a = 0$.

The deterministic model of (5.1) and (5.2) describes population average behavior and does not capture the stochasticity of the individual transitions. A type II persister would undergo random, although infrequent, switching between n and p. Randomly switching phenotype may seem inefficient, but as with bet hedging by gene noise it could be a sensible strategy for an organism that inhabits a fluctuating environment. Conditions like nutrient availability, antibiotic concentrations, or the presence of competition may occasionally take an unexpected turn, perhaps favoring the persister over other phenotypes. A clonal population may gain some benefit from having a mechanism that generates non-heritable variability. However, it is sensible to ask whether random switching is preferable to a more deliberative switching strategy. This will be discussed below.

5.2 Toxin–antitoxin systems and HipBA

Equations (5.1) and (5.2) model active switching between two distinct states, n and p. But what mechanism triggers the switching? One may plausibly ask whether persistence is really just an extreme form of heterogeneity in the growth rate λ, where the persisters make up the slow end of a broad distribution in λ. In fact molecular-level evidence supports a two-state model. Persistence is linked in many cases to so-called toxin–antitoxin (TA) systems, which occur widely in prokaryotes. TA systems typically consist of two gene products that are encoded by the same operon. Each system includes an intracellular toxin that inhibits the growth of the cell and an antitoxin that neutralizes the toxin and permits normal growth. The fact that high-persistence mutants such as the $hipA7$ strain of *E. coli* generally carry defects in TA systems suggests that their switching between persister and normal phenotypes results from competition between the toxin and antitoxin components of the system.

Many, if not most, bacterial species possess TA systems. *E. coli* has at least 10 TA systems, including the well-studied *hipBA* (which is affected in the *hipA7 E. coli* strain), *relBE*, and *mazEF*. *Mycobacterium tuberculosis* has at least 50 TA systems. The growth or dormancy of a particular cell should then reflect in some way the status of competition within each of its TA systems.

TA systems appear to function through a variety of mechanisms. While the toxin is a protein, the antitoxin may be an RNA that inhibits synthesis of the toxin, or it may also be a protein. In the first case the TA system is described as type I or type III. In the second case the TA system is called type II[4]. The *hipBA* operon controls a type II system; its toxin and antitoxin are both proteins. Its mechanism of switching is slightly complex, but as it is a classic TA system its use of feedback regulation, competition and stochasticity deserves a closer look.

The *hipBA* operon encodes two proteins, HipB and HipA, sketched in figure 5.2. HipA is the toxin. It inhibits growth of the bacterium by modifying a so-called elongation factor that is necessary for translation during protein synthesis. The antitoxin HipB neutralizes HipA by forming a tightly bound dimeric complex, denoted HipBA, with HipA. The HipBA complex in turn inhibits transcription of *hipBA* because the antitoxin has a DNA binding site that allows the complex to bind to sites on the *hipBA* promoter. Therefore the HipBA complex shuts down its own transcription.

The system has two states, subject to the activity of the operon. When active the operon produces HipB at a higher rate than HipA, generating excess HipB that neutralizes HipA and allows normal cell growth. When the operon is inactive, the HipB antitoxin tends to degrade as it is less stable than the HipA toxin. This leads to a higher relative proportion of toxin to antitoxin in the cell. Consequently toxin accumulates and growth ceases. Whether the cell exhibits normal growth or dormancy depends on the state of the operon, which is correlated with the level of antitoxin present.

The molecular competition between HipB and HipA was demonstrated at the single-cell level through measurements of the distribution of lag times for the emergence of *E. coli* from stationary phase [8]. Rotem *et al* withdrew cells from a stationary culture, spread them onto agar plates, and measured the duration of their growth arrest, i.e. the time before they resumed active growth. Generally the probability distribution for these lag times is exponential, indicating that each cell has the same probability per unit time for resuming growth. However, as shown in figure 5.2 the shape of the distribution depended on the level of HipA production, which the researchers could manipulate by adding an additional copy of the *hipA* gene. When cells made little HipA, the lag time distribution was an exponential distribution with an average rate of about 11 min: a non-growing cell switched to active growth with a probability $1/11$ min^{-1}. This may be considered normal

---

[4] These are not to be confused with type I and type II persisters.

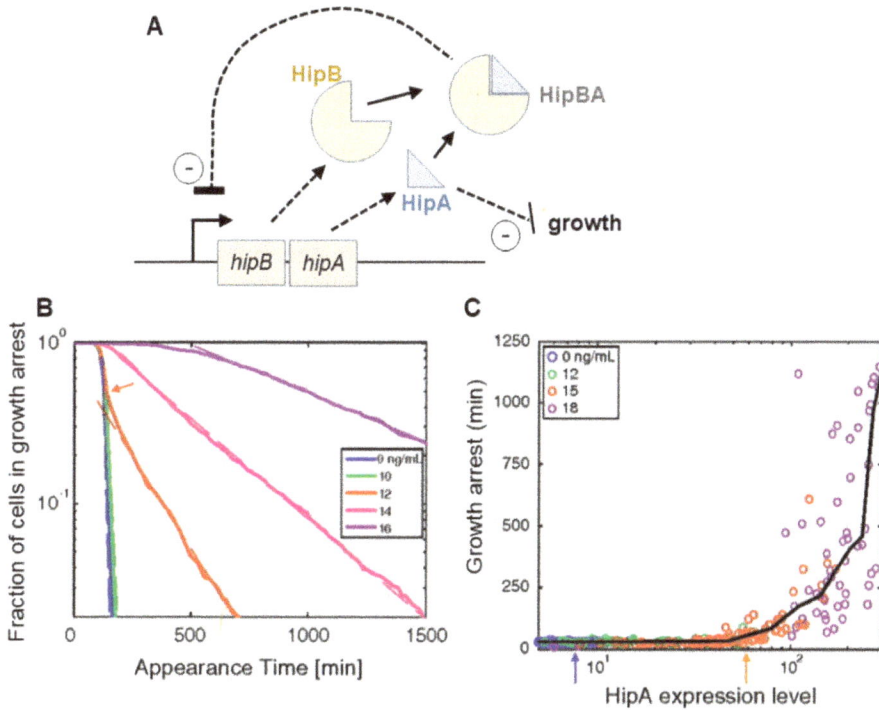

Figure 5.2. The HipBA TA system of *E. coli*. (A) HipB and HipA are products of the *hipBA* operon. HipA is a toxin that inhibits cell growth, while HipB blocks the action of HipA by binding it to form the complex HipBA. HipBA in turn inhibits transcription of the operon. As the HipA toxin is more stable than the HipB antitoxin, switching off the operon drives the cell toward dormancy. The relative concentrations of HipB and HipA determine whether the operon will remain on or off. (B) Artificially manipulating HipA levels (by introducing an additional, inducible copy of the *hipA* gene) changes the duration of the growth arrest as cells emerge from stationary phase. At the lowest HipA levels cells emerge rapidly from growth arrest, causing the number of dormant cells to fall exponentially at a rate 1/11 min^{-1} (blue curve). At the highest HipA levels the exponential has a weaker slope, indicating a rate of only 1/220 min^{-1} (purple curve). At intermediate HipA (such as the red curve) a biphasic behavior indicates that the population is divided into a fast emerging group and a slow emerging group. At these intermediate HipA levels the concentration of HipB antitoxin may or may not be sufficient to counteract the HipA toxin. The legend shows the *hipA* inducer concentration that was used. (C) Time delay before resumption of normal growth, for individual cells as a function of HipA levels (measured by a red fluorescent protein reporter). Each point shows the time at which one cell, at one HipA expression level, resumed growth. HipA expression that exceeds a threshold level (orange arrow) is likely to inhibit growth, but lower expression (blue arrow) never does. Panels (B) and (C) reproduced from [8] with permission.

behavior. When cells made large amounts of HipA, they switched far more slowly, with a rate nearer 1/220 min^{-1}. This slower rate is more characteristic of persistence. At HipA levels in between these limits, cells showed a bi-exponential lag time distribution, indicating heterogeneity in the population. The level of HipA determines the fraction of cells that emerge quickly (normal) or slowly (persister) from lag phase.

The bi-exponential distribution shows how noise in gene expression affects the competition between HipB and HipA. Because of cell-to-cell variability in HipB levels, each cell may be able to counteract a different amount of the HipA toxin. Accordingly when HipA expression is at moderate levels, the population divides into two groups where HipB is sufficient and insufficient, respectively, to counteract the toxin. By adjusting HipA levels, the experimenter can tune the system through the threshold behavior, shown in figure 5.2(C), where HipA levels begin to match HipB levels. When HipA levels are very low, all cells in the population have sufficient HipB to counteract the toxin and the growth arrest times are short. At the threshold HipA level, however, HipB levels are insufficient in some but not all cells, and so the growth delay becomes heterogeneous, with significant delays appearing in some but not all cells.

The above picture rationalizes the observation of variable growth rates. It does not explain whether the fast and slow growth phenotypes are stable states of the HipBA system. That is a question about the dynamics of the system, the feedback mechanisms and their strength. But the number of dormant cells in a growing population is miniscule, and persister cells can remain dormant for many hours. These observations suggest that transitions between normal and dormant states are rare but pronounced, and therefore that HipBA could be a bistable, stochastically switching system like those discussed earlier.

At least one model shows how bistability may arise in HipBA [7]. The model depends critically on the idea that the toxin HipA acts to slow the growth of the cell, and that this slowing of cell growth reduces the rate of dilution of proteins in the cell. In a normal growing cell the concentration of any protein is continuously being reduced by the increasing cell volume. HipA's growth-slowing action then creates a mechanism of autofeedback, enhancing its own concentration and allowing it to continue suppressing growth. The cell may then operate in two modes:

- In the first mode, the *hipBA* operon is active, producing an excess of HipB over HipA, and so HipA is detoxified. The cell therefore grows rapidly and continuously dilutes the Hip proteins, keeping the operon active, the concentration of HipA small, and the cell in normal growth.
- In the second mode, *hipBA* is inactivated by the HipBA complex, leading to cessation of HipB synthesis. HipB degradation then results in an excess of HipA, which inhibits cell growth. Slow growth prevents dilution and keeps the concentrations of HipA protein high.

Feng *et al* [7] showed that both states can be stable, at least for some parameter values. While the numbers of HipB molecules are similar in the fast and slow growth states, the two states differ in the number of HipA molecules, which is higher for the dormant (second) state. However fluctuations in HipA numbers can drive occasional, stochastic transitions between the dormant and growing states. Figure 5.3 shows simulations of such intermittent, all-or-nothing transitions.

In this model the existence of bistability depends on the cell's normal growth rate and the strength of association between HipB and HipA, as indicated in figure 5.3(B). High growth rates and stronger association tend to favor normal growth,

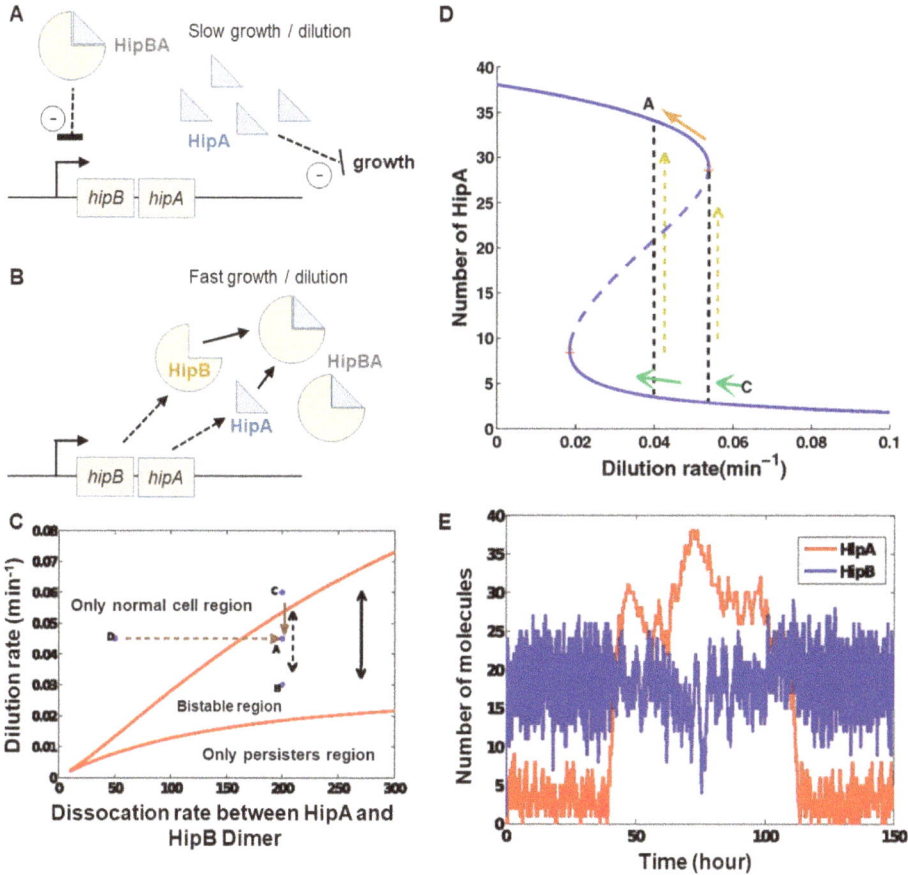

Figure 5.3. A model for bistability in the HipBA TA system [7]. (A) When transcription from the *hipBA* operon is off, degradation of HipB leads to high relative levels of HipA, suppressing growth. The slower growth prevents dilution of HipA and keeps the cell in the slow growth state. (B) When transcription from *hipBA* is on, HipB levels keep pace with HipA and counteract the toxin, allowing faster growth. The more rapid dilution of cellular protein keeps HipA concentrations low, maintaining the rapid growth state. (C) Depending on the growth or dilution rate and the binding affinity of HipA and HipB, the HipBA system may be locked on (normal), off (persistent), or bistable (intermediate region). (D) At high growth (or dilution) rates the steady state HipA level is low. As the growth rate decreases through value C, the system enters a bistable region where the HipA level can switch from low to high, triggering growth arrest. Points A and C are indicated also in (C). (E) In a computer simulation of the bistable behavior, the HipB concentration is reasonably constant over time, but HipA levels switch dramatically as the cell enters and leaves the persistent state. Panels (C)–(E) reproduced from [7] with permission.

with no dormant state possible. Slower growth rates and weaker association favor the dormant state. However, in an intermediate regime of parameters both states are possible and the system can exhibit bistability. The model potentially explains why stationary phase can induce otherwise normally growing cells to become persistent: it reduces the growth rate, stabilizing the dormant state. It also gives some insight into why the *E. coli* mutant *hipA*7 shows enhanced persistence. The allele has two point mutations that reduce the affinity of HipB and HipA, enhancing the likelihood that the system inhabits either the bistable or persistent region of the parameter space in figure 5.3(C). Models of this kind bring a richer mechanistic insight to the simple deterministic descriptions, such as (5.1), which introduced this discussion of persisters.

5.3 Bet-hedging by phenotypic switching

While the idea of bet-hedging through phenotypic switching makes sense, one may wonder if it is truly advantageous for the switching to occur randomly, as it does in (5.2). Ideally it would be possible to quantify the benefit that comes from random phenotypic switching and determine how it depends on (for example) properties of the fluctuating environment.

The idea of bet hedging in a fluctuating environment has been put on a theoretical footing by Kussell and Leibler [9]. They analyzed a model, sketched in figure 5.4, in which environmental conditions vary randomly over time, switching from one type of growth environment to another. Suppose for example that if the present condition is denoted i, then it lasts on average a time τ_i before switching to another condition j. Perhaps some environmental conditions are more likely than others, so b_{ij} represents

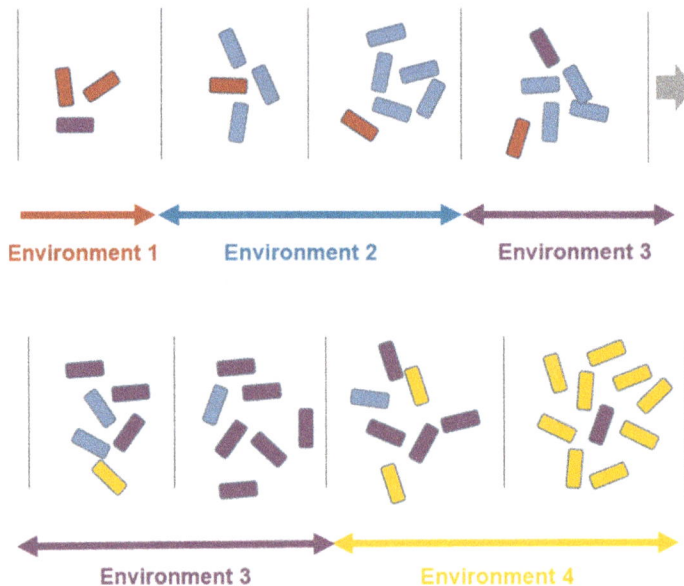

Figure 5.4. Schematic of the bet-hedging model for phenotypic switching in a fluctuating environment. Bacteria switch randomly between phenotypes, indicated by cell color. At the same time, their environment changes unpredictably over time, such that for each environment one particular phenotype enjoys faster growth. The combination of random switching by the bacteria and faster growth of the cells that have switched to the optimal phenotype leads to overall growth of the population over time. Based on Kussell and Leibler (2005) [9].

the probability that condition j will be followed by condition i. Suppose also that there is an optimal phenotype for each possible environmental condition. That is, under an environmental condition j, there is a phenotype (also identified by j) that grows faster than all the other phenotypes denoted by i, where $i \neq j$.

Then one can consider two basic strategies for phenotypic switching. The first strategy is for each cell to switch randomly between phenotypes; a cell with phenotype i may flip to any other phenotype j, via a random process, characterized by an average rate H_{ij}. Alternatively, the switching could be responsive; each cell could sense and interpret its environmental conditions in order to drive deliberate switching from its present phenotype to the optimal one.

Kussell and Leibler calculated the overall rate of growth of a population that uses one of these strategies. Their analysis showed that the random switching model can lead to an enhancement of the growth rate. In fact there is maximal growth overall, on average, when the phenotypic switching rate is $H_{ij} = b_{ij}/\tau_j$. This formula gives a recommendation for choosing H_{ij} optimally that should make sense, at least in hindsight. If the current environmental condition, j, tends to be short-lived (smaller τ_j) then it is advisable to jump out of phenotype j quickly (larger H_{ij}) and into another one. And if condition i very typically follows condition j (b_{ij} is large), then switching from j to i is statistically a good decision (larger H_{ij}). Therefore the optimum switching rate plausibly depends on b_{ij}/τ_j. Figure 5.4 illustrates the growth of a stochastically switching population in this model. The random switching strategy succeeds because all the cells tend to explore their different phenotypes, and then when one fortuitously arrives at the optimal phenotype it—and its progeny—enjoy a windfall.

The unpredictability of the environment does take a toll on growth. In calculating the overall growth rate for the stochastic switching population, Kussell and Leibler found the growth rate was diminished by an amount that depends on an interesting mathematical expression. The expression is a sum over all n possible environmental conditions and the probabilities of transitions between them,

$$I_{\mathrm{env}} = -\sum_{i,j=1}^{n} p_j b_{ij} \log b_{ij}. \tag{5.3}$$

Here p_j is the probability of environment condition j. This term will have a larger (negative) impact on growth when all the transition probabilities b_{ij} are of similar magnitude. It will have smaller impact when a few of the transitions are highly probable and the rest are very improbable. This sum measures the unpredictability of the environment and it is referred to as the information content of the environment.

We will address the measurement of information in more detail in the discussion of bacterial communication. Here, rather than rationalize (5.3) in detail, it is more useful simply to note that it quantifies the unpredictability of a statistical quantity (the fluctuating environment) as a sum over the probabilities of different values of that quantity. Describing an unpredictable environment as having 'high information' is justified by the fact that one learns more from measuring an unpredictable variable than from measuring a predictable one. If the switching behavior of the environment is completely predictable, then one learns nothing from measuring it and $I_{env} = 0$. If the environment has a very large number of possible behaviors, of nearly equal probability, then one learns a great deal from measuring the environment and so I_{env} is much larger. Large I_{env} makes it more difficult for the organism to switch successfully by a random process, and so it limits growth.

Kussell and Leibler also considered the alternative strategy where cells sense and interpret their environment and then switch phenotype accordingly. Within the framework of the model as described above, one can suppose that the rate of transition from phenotype j to i depends on the current environmental condition. Ideally the organism does a perfect job of sensing the environment, so if the current condition is k, then all the rates for switching phenotypes $j \rightarrow i$ are zero unless $i = k$, and the switching rate in that case should be quick compared with the duration τ_k of the condition. This approach does lead to a higher growth rate, except that measurement of the environment has to come at a metabolic cost to the organism. Kussell and Leibler modeled this by imposing a cost or 'tax' c on the total growth rate in the presence of responsive switching. Responsive switching is advantageous over stochastic switching only if the metabolic cost c is not too large. Perhaps not surprisingly, the maximum permissible metabolic cost depends on the information content, equation (5.3), of the environment. The larger I_{env}, the more unpredictable the environment, and so the higher the cost c the organism can pay and still benefit from responsive switching. Whether random phenotypic switching is an efficient strategy depends on the degree of unpredictability of the environment, weighed against the metabolic cost of implementing a more intelligent method. Of course a cell can combine both approaches. As will be discussed later, there are bacterial phenotypic switches that employ both an element of randomness and significant input received from environmental cues.

References

[1] Davies J and Davies D 2010 Origins and evolution of antibiotic resistance *Microbiol. Mol. Biol. Rev.* **74** 417–33

[2] Brauner A, Fridman O, Gefen O and Balaban N Q 2016 Distinguishing between resistance, tolerance and persistence to antibiotic treatment *Nat. Rev. Microbiol.* **143** 14320–30

[3] Bigger J W 1944 The bactericidal action of penicillin on *Staphylococcus pyogenes Irish J. Med. Sci.* **19** 585–95

[4] Balaban N Q, Merrin J, Chait R, Kowalik L and Leibler S 2004 Bacterial persistence as a phenotypic switch *Science* **305** 1622–5

[5] Moyed H S and Bertrand K P 1983 *hipA*, a newly recognized gene of *Escherichia coli K-12* that affects frequency of persistence after inhibition of murein synthesis *J. Bacteriol.* **155** 768–75 PMCID: PMC217749

[6] Gefen O and Balaban N Q 2009 The importance of being persistent: heterogeneity of bacterial populations under antibiotic stress *FEMS Microbiol. Rev.* **33** 704–17

[7] Feng J, Kessler D A, Ben-Jacob E and Levine H 2014 Growth feedback as a basis for persister bistability *Proc. Natl. Acad. Sci.* **111** 544–9

[8] Rotem E, Loinger A, Ronin I, Levin-Reisman I, Gabay C, Shoresh N, Biham O and Balaban N Q 2010 Regulation of phenotypic variability by a threshold-based mechanism underlies bacterial persistence *Proc. Natl. Acad. Sci.* **107** 12541–6

[9] Kussell E and Leibler S 2005 Phenotypic diversity, population growth, and information in fluctuating environments *Science* **309** 2075

The Physical Microbe
An introduction to noise, control, and communication in the prokaryotic cell
Stephen J Hagen

Chapter 6

Communication

One obvious difference between microbes and higher organisms is in the size of the fundamental viable unit: a solitary microbial cell is all that is needed to generate a population. But the fact that one cell comprises in many respects a complete package does not mean that microbes are solitary organisms. They have extensive social behavior. They inhabit both single-species and mixed-species communities in which they interact, cooperate, compete, and communicate with each other. Social behavior is particularly important where bacteria grow to high population densities, as occurs in the human digestive tract, which is inhabited by numerous bacterial species that maintain a dynamic balance that is closely tied to the health of the host. Bacterial social behavior is important in biofilms, the microns-thick plaques that often form on surfaces in contact with nutrient-rich media. Biofilms are highly structured communities of bacterial cells, often a mixture of species, embedded in a matrix of their own secretions—proteins, polysaccharides, and other compounds. The slimy layers that form on wet surfaces like rocks in creek beds, medical devices like catheters, and even human teeth (dental plaque) are all examples of bacterial biofilms[1]. Biofilm bacteria engage in both interspecies and intraspecies signaling, through mechanisms discussed below, in order to coordinate a range of behaviors.

Given that microbes have need to communicate, one may ask what physical modes of communication are available to them. Some mechanisms that serve nicely at the macroscopic scale are probably ineffective at the single-cell scale. Signaling with light is not strictly impossible—some microbes emit light, and some respond to light[2]—but the evidence that microbes exchange information through light is not yet overwhelming [1, 2]. It would be difficult for microbes to generate vibrations, and so communication by sound would likely be unwieldy. Chemical signaling however is

---

[1] The tangible sensation of slime on the teeth that follows consumption of a sugar-rich beverage is due to the rapid growth of biofilm by sucrose-loving oral bacteria such as *Streptococcus mutans*.

[2] *Vibrio fischeri* and *Vibrio harveyi* are famously bioluminescent. Photosynthetic bacteria have a circadian cycle that synchronizes with daylight.

ubiquitous in biology. It works fine at the cellular level. Chemical signaling is consequently the best understood and (probably) most widespread mechanism of bacterial communication. In addition, an increasing number of studies show evidence of electrical communication among bacteria, particularly in biofilms. There is also recent evidence of contact (touch) communication, which we will not explore here. This section will examine the physical mechanisms of electrical and chemical communication in bacteria.

6.1 Chemical communication

The paradigm of bacterial communication has long been *Vibrio fischeri*, a bioluminescent marine bacterium. *V. fischeri* generates light through a bioluminescence reaction[3] that is catalyzed by the bacterial luciferase enzyme. It may seem mysterious that solitary cells present at low density in seawater would generate light. However, *V. fischeri* grows to high population densities when it enters a symbiosis with certain fish and squid species, such as the squid *Euprymna scolopes*. The squid possesses a specialized (and very hospitable to *V. fischeri*) cavity known as the light organ. Inside the light organ *V. fischeri* grows to densities of 10^9–10^{10} cells cm^{-3}, concentrating a large number of weakly luminescent individual bacteria into one brightly luminous colony.

Light production serves a symbiotic purpose, as it can allow the host animal to evade predators or to lure prey. But a solitary *V. fischeri* generating light in the open ocean is evidently wasting energy. In its fully bioluminescent state, one *V. fischeri* may emit 10^3–10^4 photons/s at a wavelength near 490 nm. This drains power at a rate as high as 4 fW per cell. For a cell with a mass of 0.6 pg (table 1.1) this equates to 7 W kg^{-1}, likely a substantial burden on metabolism.

Presumably to avoid bioluminescing when not in the light organ, *V. fischeri* regulates its light emission through a mechanism that is sensitive to its own population density. As shown in figure 6.1, each cell produces and secretes a small soluble molecule, *N*-3-oxo-hexanoyl homoserine lactone (3OC6 HSL), that diffuses across the cell membrane and accumulates in the environment. At the same time, each cell detects the environmental concentration of 3OC6 HSL. When the concentration reaches a threshold value, a transcriptional regulatory circuit activates expression of the so-called *lux* genes, which produce the bioluminescence. Therefore bioluminescence is triggered when the *V. fischeri* population density has reached a sufficiently high density. Light production is a cooperative behavior that is regulated at the population level by an exchange of the chemical signal, 3OC6 HSL.

Although the details get complicated (see below), two proteins, LuxI and LuxR, form the core of the *V. fischeri* signal production and detection mechanism, shown in figure 6.1. The enzyme LuxI synthesizes the 3OC6 HSL signal, while the receptor LuxR binds the 3OC6 HSL to form a transcriptional activator complex. This complex targets a DNA region known as the *lux* box to stimulate transcription of

---

[3] The bioluminescence is generated by an enzyme-driven chemoluminescent reaction. The enzyme luciferase uses O_2 to oxidize the biomolecule FMNH$_2$ and a long-chain aldehyde.

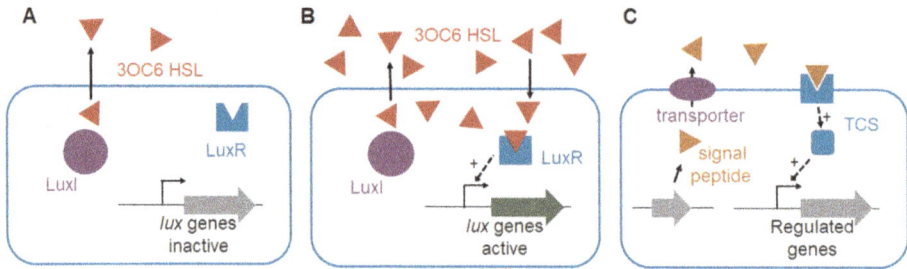

Figure 6.1. Chemical communication in bacteria. The LuxIR system of *V. fischeri* is a classic Gram-negative quorum communication system. It provides a switching mechanism by which the organism can regulate the expression of its *lux* genes in response to its own population density. (A) LuxI produces the 3OC6 HSL signal molecule, which diffuses out of the cell and accumulates in the extracellular environment. At low bacterial populations the environmental concentration of 3OC6 HSL is small, LuxR remains inactivated, and so the *lux* genes are switched off. (B) As the bacterial population rises, the accumulating 3OC6 HSL binds to the intracellular receptor LuxR to form a transcriptional activator for the *lux* genes. The LuxR-HSL complex switches on these genes, leading to a bright bioluminescence. (C) Quorum sensing in Gram-positive species is a similar behavior, implemented by a different mechanism. The signal molecule is typically a small peptide that must be exported from the cell by a dedicated transporter system. The accumulated extracellular signal in turn is detected by a TCSTS. The TCSTS has two parts: (1) a membrane-localized receptor molecule, upon detecting the signal, activates (2) an intracellular molecule—the response regulator—that triggers transcription of the regulated genes.

the seven-gene *lux* operon (*luxICDABEG*). The *lux* operon includes *luxI* as well as the genes that encode the enzyme luciferase (LuxA and LuxB) and the other components required for bioluminescence.

Because of its clear link to population density in both *V. fischeri* and the closely related (and also bioluminescent) *Vibrio harveyi*, this chemical communication behavior is often called quorum sensing. Quorum sensing turns out to be a common behavior in bacteria. Numerous species employ communication mechanisms that are analogous to LuxI/LuxR, in order to regulate a variety of behaviors. These behaviors range from bioluminescence to aggregation, forms of motility, production of surfactants and secretions, biofilm formation, sporulation and genetic competence[4], as well as various aspects of virulence in pathogenic bacteria. Gram-negative species commonly signal using HSL signals[5], although not necessarily the 3OC6 HSL used by *V. fischeri*. Different organisms employ different HSLs (differing in their carbon chains), and each sensing system generally detects and respond to its own particular HSLs with great specificity. HSLs activate signaling circuits by acting as inducers, binding to transcription factors. Gram-positive bacteria also

---

[4] Sporulation, mentioned earlier, is a form of division that generates a dormant daughter cell. Genetic competence is the ability to take up and incorporate DNA from the environment.

[5] More precisely, the signals used by Gram-negatives are *N*-acyl homoserine lactones. An HSL consists of a homoserine lactone ring and an acyl (R-CO) side chain, 4–16 carbon atoms in length. An HSL is identified by the length of its side chain and by the presence of any carbonyl or hydroxyl groups on that chain. For example, 3OC6 HSL has a six carbon side chain with a carbonyl group on the third carbon. C8 HSL has a simple chain of eight carbons.

engage in quorum sensing, using small peptide signals instead of HSLs (figure 6.1). These peptides drive phosphorylation pathways, such as two component signal transduction system (TCSTSs), rather than interacting directly with transcription factors.

HSLs and other chemical signals are examples of pheromones, or molecules that are exchanged by plants and animals (including trees and insects) in order to communicate. Individuals exchange pheromones in order to trigger behavioral responses from each other. In fact, in order to emphasize this larger biological context[6], quorum sensing is sometimes called pheromone sensing, or just chemical communication. Nevertheless, it has generated broad interest not only in microbiology, but also in fields such as synthetic biology, applied mathematics, and engineering. For example, the synthetic biology community has taken interest because regulatory modules such as LuxIR can be modified and spliced into different organisms, allowing researchers to design new interspecies and intraspecies interactions, or to explore ideas such as bacteria-powered computers. Moreover, the idea that single-celled organisms negotiate group behaviors by engaging in chemical 'chitchat' is charming and has inspired a large number of cleverly titled review articles [3]. The colorful story of the bioluminescent *V. fischeri* and its symbiotic niche serves as an appealing illustration of why bacteria might need to sense their own population or 'quorum'.

However, it is probably a mistake to overemphasize the population counting role of pheromone signaling systems and overlook their other possible functions. Even the seemingly classic quorum-sensing pathway of *V. fischeri* regulates multiple behaviors in addition to bioluminescence and, as shown in figure 6.2 has complexity significantly beyond the core LuxIR system. As shown schematically in figure 6.2 the system has two additional pheromone-sensing pathways in addition to LuxIR: 3OC6 HSL is produced and detected by the LuxIR system, C8 HSL is a separate signal that is produced and detected by AinS and AinR, respectively, and a third compound known as AI2 signals through an additional pathway involving LuxS and LuxQ. In fact AI2 appears to be an interspecies signal[7], synthesized and recognized by numerous other species including Gram-positives. The use of three pheromones, including the interspecies signal AI2, suggests that this system is not just counting population density. Further, the regulatory pathways controlled by quorum-sensing systems such as LuxIR very often respond to a number of additional environmental cues, such as chemical conditions of pH and oxygen concentration, in addition to HSL concentration. Many quorum-controlled regulatory networks employ positive feedback loops, leading to nonlinearity and bistability, and exhibit a high degree of noise.

If an organism needs only to detect its population density, it is not clear why its chemical signaling circuit should incorporate all of these features. Accordingly there has been ample discussion in the literature about the kinds of environmental information such systems can gather and what they can do with that information,

---

[6] ...and perhaps also to avoid an overly narrow interpretation of the behavior, as discussed below.

[7] AI2 is neither a peptide nor an HSL. It is a furanosyl borate diester, chemically dissimilar.

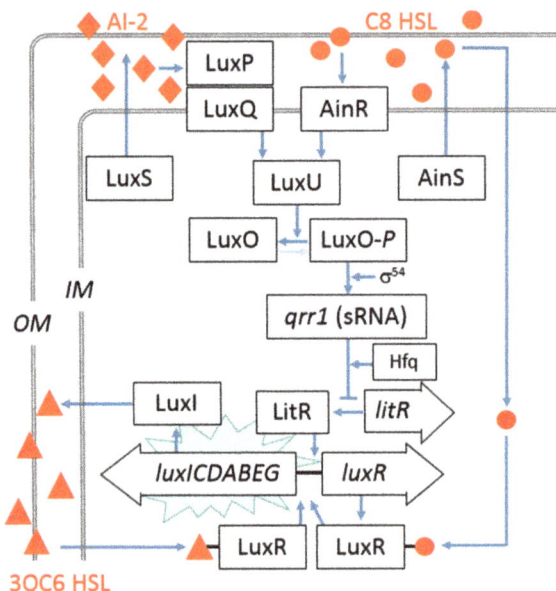

Figure 6.2. Although the LuxIR system of *V. fischeri* is the classic example of a quorum-sensing circuit, it comprises only a portion of the chemical signaling network that controls bioluminescence. As shown in this very schematic representation, the network appears more complex than is strictly necessary to detect population density alone. The organism synthesizes and detects three different diffusible signals (3OC6 HSL, C8 HSL, and AI2), using multiple pathways and layers of control, with crosstalk between signal channels and elements of positive feedback. The three enzymes LuxI, AinS, and LuxS synthesize the signals, which are released to the extracellular environment beyond the outer membrane (OM). Other cells detect these signals using LuxR, AinR, and LuxPQ, respectively. C8 HSL and AI2 drive a shared phosphorylation cascade involving LuxU and LuxO as well as the small regulatory RNA *qrr1*. Signaling ultimately converges at the *lux* operon, *luxICDABEG*, which controls bioluminescence among other behaviors. Note that the *lux* operon includes the 3OC6 HSL synthase LuxI, leading to a positive feedback loop in 3OC6 HSL synthesis. Adapted from [4].

particularly at the individual cell level. For example it seems highly likely that bacteria can use these circuits for interspecies (in addition to intraspecies) communication. *V. harveyi*, discussed below, produces and detects the AI2 signal that is also used by numerous other Gram-negative and Gram-positive species. AI2 may be a sort of universal signal that allows each species to sense the presence of other species and gauge its own role in the total microbial community. It is worthwhile to explore from a physical perspective some of the limitations on the information that such circuits can deliver to the cell.

6.1.1 Alternative interpretations of quorum sensing

It has been suggested that quorum-sensing circuits need not serve a social or communication function at all. Perhaps such a circuit can provide other kinds of information to a solitary cell. Suppose for example that a cell releases a signal molecule, and the local concentration of the signal eventually reaches the threshold that triggers a sensing pathway. If the cell is a member of a population in a well-mixed environment, activation of the pathway indicates that the population has reached a critical size. However, if the cell is solitary and inhabits a small confining volume, then activation of the circuit may inform the cell about the effective volume of its environment; the accessible volume must equal the threshold signal concentration divided by the total amount of signal produced. Alternatively, if the cell is in a poorly mixed environment, such as a biofilm or an aggregate, activation could inform the cell that local diffusion or advection are inefficient, so that cellular secretions in general will not travel far in the local environment. For solitary cells that secrete enzymes into their environment in order to digest or capture nearby nutrients, such information about physical confinement or inefficient diffusion is valuable. In short, quorum-sensing pathways could in principle detect either population (quorum), confinement, or the local efficiency of transport. There are also schemes (discussed later) where quorum signals could control the rate at which a behavior switches on.

This perspective has raised questions of whether quorum-sensing behavior should instead be interpreted as diffusion sensing or environment sensing [5]. However, while these distinctions seem physically straightforward they become murkier when placed in more realistic biological context. For example, on the time scale over which signal accumulates and quorum sensing activates, a solitary bacterium confined to a small volume could grow into a cluster of many clonal cells. Then the distinction between one solitary cell and multiple communicating cells is not so clear. Second, it may not make sense to ask whether the pheromone circuit serves the individual microbe or the population, because natural selection does not distinguish between these cases [6]. Natural selection maximizes the fitness of a certain genetic program, and if a population consists of close relatives who share that program, then the two cases are not distinguishable. In any case, while experiments do show that quorum circuits can activate when single cells inhabit small volumes, this does not prove the 'intent' of those circuits. Chemical signals can regulate behavior in a useful way under a variety of population and environment conditions. Cells function in complex environments that may be chemically and physically heterogeneous, with intercellular cooperation, competition, and diffusion all playing a role. The complexity of the circuits reflects the need to optimize information gathering and processing in such confusing environments.

Exercise (Diffusion sensing) *Suppose that n_0 bacterial cells initially inhabit a volume V. The cells divide at a rate λ while each cell secretes diffusible signal into its environment at rate k (moles/time). After a time t, the signal concentration has reached c_0, triggering a regulatory pathway. Find the relationship between the time t and the volume V. Does t provide a sensitive measure of V?*

6.1.2 Sensitivity of chemical communication

Setting aside the quandary of distinguishing 'group' from 'individual' benefits in bacterial signaling, one can still pose the more focused physical question of how accurately or precisely an individual cell can gather information through chemical communication. For example, if microbes can measure diffusing signals released by other microbes, then perhaps individual cells can signal to each other over relatively long distances. If a few sophisticated cells employed multiple signals with different diffusion coefficients, perhaps they could even use the local signal concentrations to deduce their locations with respect to each other. Then chemical signaling would allow a sort of position-sensing or triangulation behavior. Does chemical communication allow this kind of precision? Some aspects of this problem have been understood for a very long time; more recent single-cell data demonstrate that gene noise is too important a part of the story to ignore. Here we consider some of the physical limitations to the sensing of chemical signals by individual microbes.

One might guess that if a chemical signal is present at low (such as nanomolar) concentration in the environment, then cells may have difficulty counting the signal molecules and assessing the concentration accurately. However, as long as the cell has a tool, perhaps a set of dedicated receptors, for counting diffusing particles, it can accurately gauge signal concentration. This was famously shown by Berg and Purcell [7] as follows. Suppose that a diffusible signal is present at true[8] concentration $\bar{c}$ (molecules cm^{-3}). For the cell to make an estimate, denoted c, of the concentration, it must count the number n of copies of the molecule that are present within some small volume v. If the signal enters the cell, that detection volume might be the volume of the cell itself. If for example the signal is present at $\bar{c} = 1$ nM and the cell has $v = 1$ μm^3, then the average or expected result for n will be $\langle n \rangle = \bar{c}v = 0.6$ molecules. Any individual measurement of n allows an estimate $c = n/v$, although the accuracy of this estimate is compromised by statistical fluctuations in n.

How large are the fluctuations of n and c? As n is subject to Poisson statistics its variance is $\sigma_n^2 = \langle n \rangle$, so the relative uncertainty in the cell's measurement of n is

$$\frac{\sigma_n}{\langle n \rangle} = \frac{\sqrt{\langle n \rangle}}{\langle n \rangle} = \frac{1}{\sqrt{\langle n \rangle}} = \frac{1}{\sqrt{\bar{c}v}}. \tag{6.1}$$

Because $\sigma_c^2 = \sigma_n^2/v^2$, the relative uncertainty in the cell's estimate for c is the same as that in n, $\sigma_c/c = 1/\sqrt{\bar{c}v}$. In our example that comes to $(0.6)^{-1/2} = 1.29$, a large relative uncertainty.

The cell can improve its estimate of c by measuring n repeatedly and averaging the results together. This requires allowing diffusion to take its course, replacing one group of molecules with other, statistically independent groups over a period of time. If the molecule has a diffusion coefficient D (cm^2 s^{-1}) and the volume v has width (roughly) $a = \sqrt[3]{v}$, then during a time $\tau = a^2/D = v^{2/3}/D$ most of the molecules initially present within v will diffuse away and be replaced by a new sample. To make N essentially independent measurements of n the cell needs to measure the average n over an interval $t = N\tau$. Averaging a Poisson-distributed quantity N times reduces the relative uncertainty by a factor $\sqrt{N}$. Therefore after a time t the relative uncertainty in the cell's estimate of c is

---

[8] Specifically, $\bar{c}$ is the time-averaged concentration at the cell's location.

$$\frac{\sigma_c}{c} = \frac{\sigma_n}{\langle n \rangle} \frac{1}{\sqrt{N}} = \frac{1}{\sqrt{N \bar{c} v}}. \tag{6.2}$$

With $N = tD/a^2$ and $a = v^{1/3}$ this can be expressed in terms of the measurement interval t,

$$\frac{\sigma_c}{c} = \frac{1}{\sqrt{\bar{c} D t}} \frac{1}{v^{1/6}}. \tag{6.3}$$

Measurement precision improves gradually as the integration time t increases. For a typical small molecule with $D = 10^{-5}$ cm^2 s^{-1}, $\bar{c} = 1$ nM, and with $a \simeq 1$ μm, the uncertainty can be reduced below 1% if the cell can just integrate for a rather brief $t = 16$–20 s. Because the time scale for activating a transcriptional regulatory pathway is measured in minutes not seconds, it appears that ample integration time is built into the system[9]. Low concentrations of HSL are not an excuse for poor measurement precision by a chemical communication system such as *lux*.

Accordingly many pheromone systems respond well at nanomolar signal levels. Figure 6.3 shows the bioluminescence of *V. fischeri* as a function of the 30C6 HSL pheromone concentration. When the cells are placed in an environment that provides a fixed and stable concentration c of the pheromone, the relative luminescence L of the overall population follows a Hill function

$$L(c) = \frac{c^m}{c^m + K^m}, \tag{6.4}$$

where the Hill coefficient is $m \simeq 2.6$ and $K \simeq 200$ nM [8].

This argument only demonstrates that bacteria can on average detect a signal at low concentration. It does not address the precision of the transcriptional regulatory pathway that interprets the signal. These pathways are subject to noise in gene expression, such that fluctuations in the output can be substantial in comparison to the total dynamic range. Consequently, at the level of the individual cells the precision of output may be surprisingly poor. Figure 6.3 shows that the light output from individual *V. fischeri* varies over an order of magnitude, with a coefficient of variation near 100%. Even at micromolar pheromone concentrations that completely saturate the average light output from the population, many individuals emit little or no light. And although the HSL concentration is held constant, the brightness of individual cells fluctuates drastically over time scales of 30–60 min. Therefore the input/output function (6.4) applies at the population level, not at the single-cell level. Small changes in signal input appear to have little if any consequence for the light output from the individual cell. Signal information is lost within the transcriptional pathway.

---

[9] Signal transduction pathways, such as phosphorylation cascades can operate much more rapidly, however. The outputs of these pathways may be far more sensitive to the temporal limitations in measuring c. This is the case with bacterial chemotaxis, where the bacterium needs to swim quickly toward its food, not remain in place and wait for an accurate measurement of the concentration.

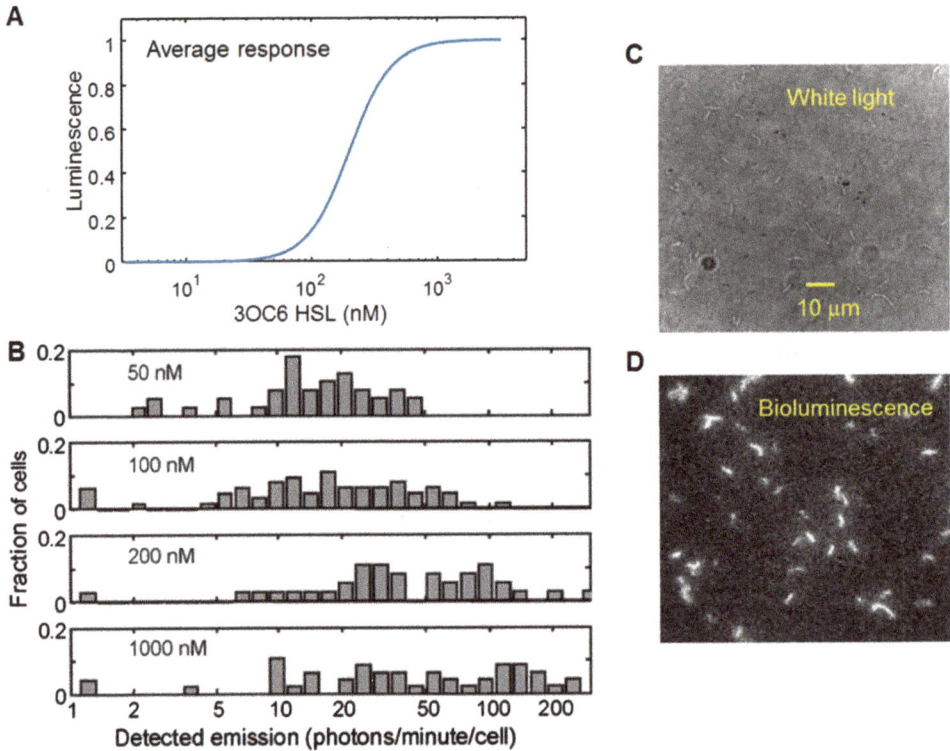

Figure 6.3. (A) The total luminescence of a population of *V. fischeri* rises when the environmental concentration of the pheromone 3OC6 HSL is above roughly 50 nM. The population response saturates at signal concentrations above roughly 1 μM. (B) At the level of individual cells, however, there is great variability in light output, regardless of signal concentration. Even when 3OC6 HSL is present at levels near 1 μM, individual cells within the population may differ more than tenfold in their light emission. (C)–(D) A field of individual *V. fischeri* is imaged under (C) external (dark field) illumination and (D) their own bioluminescent emission. Some cells luminescence brightly while others are virtually dark. Reproduced from [8].

These data raise the question of how well the individual cell processes signal information. One can phrase the question as, how much information is transmitted through the transcriptional pathway? Alternately, how many distinct levels of HSL concentration can the LuxIR circuit of one cell resolve? These questions can be answered precisely using the definitions of information and mutual information, as discussed below.

6.1.3 Information in chemical communication

In the discussion of persister phenotypes, the growth of a population was impeded not as much by the variability as by the unpredictability of the fluctuating environment. Kussell and Leibler [9] found that the growth rate was reduced by an amount that is directly proportional to the information I_{env} of the environment, defined in (5.3). The information in the environment is one application of the more general, mathematical definition of information that is useful in the analysis of all types of communication systems. A brief explanation of this definition follows.

Suppose that x is a measurable quantity, such as the concentration of signal, and X is a result obtained in one particular measurement of x. If measurements of x always produce the same X, then x is completely predictable and the measurements are informative. One gains zero information by measuring x. On the other hand, if many different X can be obtained, then a measurement of x does yield information. If for example four different X are equally likely to occur, then measuring X is as informative as learning the value of a (previously unknown) number between 1 and 4. As a two-bit binary number has four possible values, one can say that measuring x in this case is like being told a two-bit binary number: one learns two bits' worth of information.

In real measurement situations not all values are equally likely. If certain X occur more frequently than others, the measurement results can be guessed or anticipated to some extent. Predictability makes the measurements, on average, slightly less informative. Measurements are more informative if their results X tend to be more equally distributed across their different possible values.

All of this implies that the information present in x, or gained by measuring it, depends on the number of different values that X can take and the probability distribution $P(X)$ for those values. Claude Shannon presented a formal definition of information that captures these and other sensible properties. The information in x is

$$I(x) = -\sum_X P(X)\log_2 P(X). \tag{6.5}$$

The sum ranges over all possible values of X. The base-two logarithm in the sum means that information is measured in bits: if eight possible values X can occur, with equal probability ($P = 1/8$ for each value X), then $I = \log_2 8 = 3$ bits, because each measurement of x is as informative as learning a three-bit binary number. Shannon's definition has many appealing properties, which one can enjoy by reading Shannon's classic article [10]. Equation (6.5) can be applied to all types of measurables and signals, regardless of the mechanics of how the information is being delivered. The amount of information in a signal is the extent to which a measurement of that signal reduces one's ignorance.

With bacterial communication, the question is how much information flows from an input x (the signal concentration) to an output y (gene expression). In an ideal, noiseless signal pathway, each input value X would lead predictably to one output Y. Measuring the output would be a perfectly good way to discover the input, and vice versa. This ideal, predictable behavior is represented by the (average) input/output function of *V. fischeri*, $L(c)$ in equation (6.4). But if there is noisy gene expression in the pathway, the output L is an imperfect indicator of the input c. One could attempt to characterize the information flow through the noisy pathway in terms of a quantity like the correlation coefficient between x and y. However this is unreliable, as the correlation can be zero for some input/output relationships even if noise is absent. A more robust approach, less dependent on the details, is to use (6.5) to find

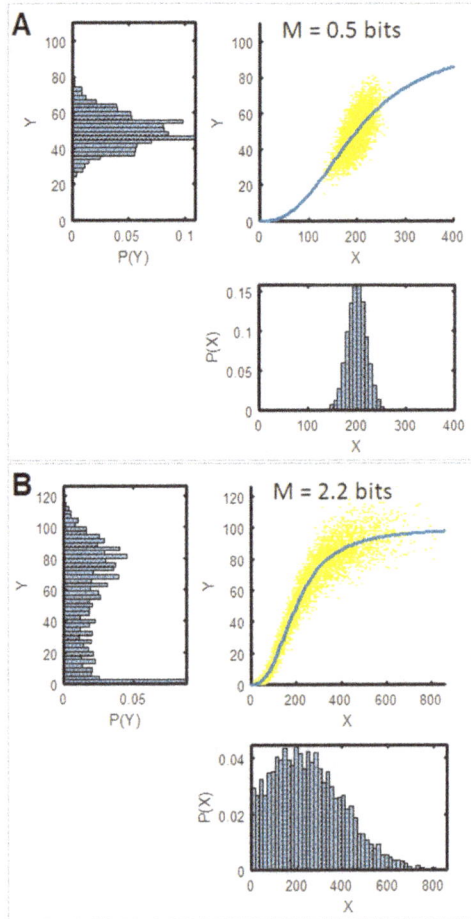

Figure 6.4. If a signal X is processed to generate an output Y, the mutual information between X and Y depends on the noise in the processing circuit and on the probability distributions $P(X)$ and $P(Y)$. In this computer-generated example, a variable X (such as the environmental pheromone concentration) is detected by a regulatory pathway that produces an output Y (such as the activation of an operon like *lux*). The solid blue curve in each group shows the deterministic X–Y behavior analogous to (6.4). However, the pathway adds noise, so the actual value of Y for a given X is noisy as shown by the yellow dots. The histograms for $P(X)$ and $P(Y)$ are shown at the bottom and left side, respectively, for each group. (The blue, deterministic curve is the same Hill function, like that of (3.9), in each case. The noise in Y is generated by a Poisson distribution whose mean is the deterministic value of Y.) (A) If $P(X)$ is concentrated around a narrow range of X that does not elicit the full range of outputs, the mutual information $M(x, y)$ between X and Y (6.6) is only 0.5 bits. (B) if $P(X)$ is broad enough to stimulate the full range of outputs, then the mutual information is significantly higher, $M(x, y) = 2.2$ bits. (C) If $P(X)$ is so broad that the pathway is almost always activated then $M(x, y) = 1.6$ bits.

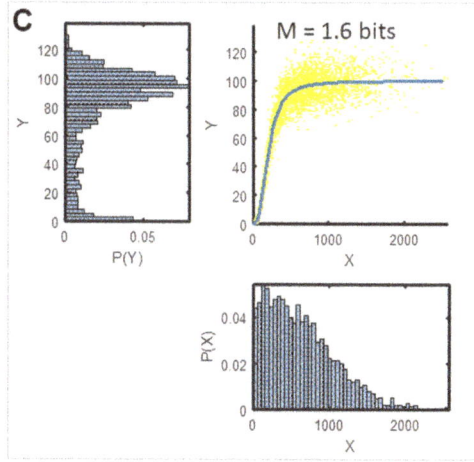

Figure 6.4. (Continued.)

how much one learns about the input value X by measuring the output Y. Specifically, how much does a measurement of Y reduce one's ignorance of X (or vice versa)?

As X and Y will both be unpredictable to some extent, they can be characterized by probability distributions $P(X)$ and $P(Y)$. Then the mutual information $I(x, y)$ is a convenient metric that describes the amount of information that one acquires about x by measuring y:

$$M(x, y) = \sum_X \sum_Y P(X, Y)\log_2 \frac{P(X, Y)}{P(X)P(Y)}. \tag{6.6}$$

Here X and Y are possible values of x and y, respectively. $P(X)$ and $P(Y)$ are the probabilities of observing values X and Y, while $P(X, Y)$ is the joint probability that a particular result (X, Y) is observed. Note that the mutual information is symmetric in the sense that one learns as much about x from measuring y as the reverse. Also note that if the variables X and Y are completely uncorrelated, so that one has no bearing on the other, then $P(X, Y) = P(X)P(Y)$ and $M(x, y) = 0^{10}$.

Three elements affect the mutual information between x and y. The first is the ideal (noiseless) input/output relation between the two variables, which affects $P(X, Y)$. The second is the amount of noise in the conversion from X to Y. The third is the probability distribution $P(X)$ of the input signal. If one were designing a signal-sensing circuit one might hope to choose all three of these elements in a way that optimizes the mutual information. The simulation in figure 6.4 illustrates how

---

[10] Equation (6.6) assumes that X and Y are both discrete variables. If they are continuous variables the sums can be rewritten as integrals.

the mutual information in a circuit analogous to that of *V. fischeri* changes as the probability distribution for the input signal is modified. Although on average there is a deterministic relationship (such as $L(c)$ in (6.4)) in which x activates y, the output is somewhat noisy (as modeled in the figure by a Poisson distribution) and the input signal may vary over a limited range. Both of these features can reduce the mutual information between x and y. The effect of noise is reasonably obvious, but it is worth noting how the mutual information is affected by the distribution $P(X)$. If, as in figure 6.4(A), X rarely deviates far from its usual value, so that $P(X)$ is sharply peaked, then Y is mostly just noise and contains almost no information about what is happening at the input. On the other hand, there is only a limited range of values X for which Y is very sensitive to X. Therefore if X has an exceptionally broad distribution, as in figure 6.4(B), then it is often outside the useful range, Y is likely to be sharply peaked around one value, and the mutual information is reduced.

Note in the simulation that M is no more than one or two bits. Despite the fact that the deterministic curve has a large vertical range, no more than 2–4 different values of x are meaningfully distinguished by measurement of y. One may wonder if this is characteristic of biological signaling pathways. It seems plausible that the input/output function implemented by a natural transcriptional regulatory network is likely to be the one that optimizes the mutual information, given the inevitable noise levels and the probability distribution for the input signal concentration in the organism's environment. In fact one may calculate, for a regulatory pathway where the input/output function and the noise levels are known, an input probability distribution that maximizes the mutual information[11]. Generally the mutual information takes its maximum value when the distribution $P(X)$ just spans the rising edge of the response curve, as occurs in figure 6.4B. This optimum design seems to occur in at least some eukaryotic signaling systems, as described in [11], where there may be about 1.7 bits of mutual information between the input signal concentration and a transcription level that is directly stimulated by that signal. The state of the output allows one to detect slightly more than three distinct states of the input.

The performance of microbial signaling systems seems to be worse than this. Figure 6.3 certainly suggests that the bioluminescence of *V. fischeri* contains very poor information about the HSL concentration. Actually calculating the mutual information for a real bacterial communication system is slightly tricky as $M(x, y)$ depends on the probability distribution $P(X)$ for the HSL signal concentration. One does not know all the concentrations the cell encounters during the relevant stages of its life in the natural environment. In order to estimate mutual information in such systems one has to make some assumptions about what signal levels are likely.

V. harveyi is a relative of *V. fischeri* that regulates bioluminescence using the pheromone-sensing system shown schematically in figure 6.5. The system is somewhat different from that of *V. fischeri* as the input from three separate signals (AI1,

---

[11] Alternatively the cell could maximize M by optimizing the signal concentrations to match the input/output function.

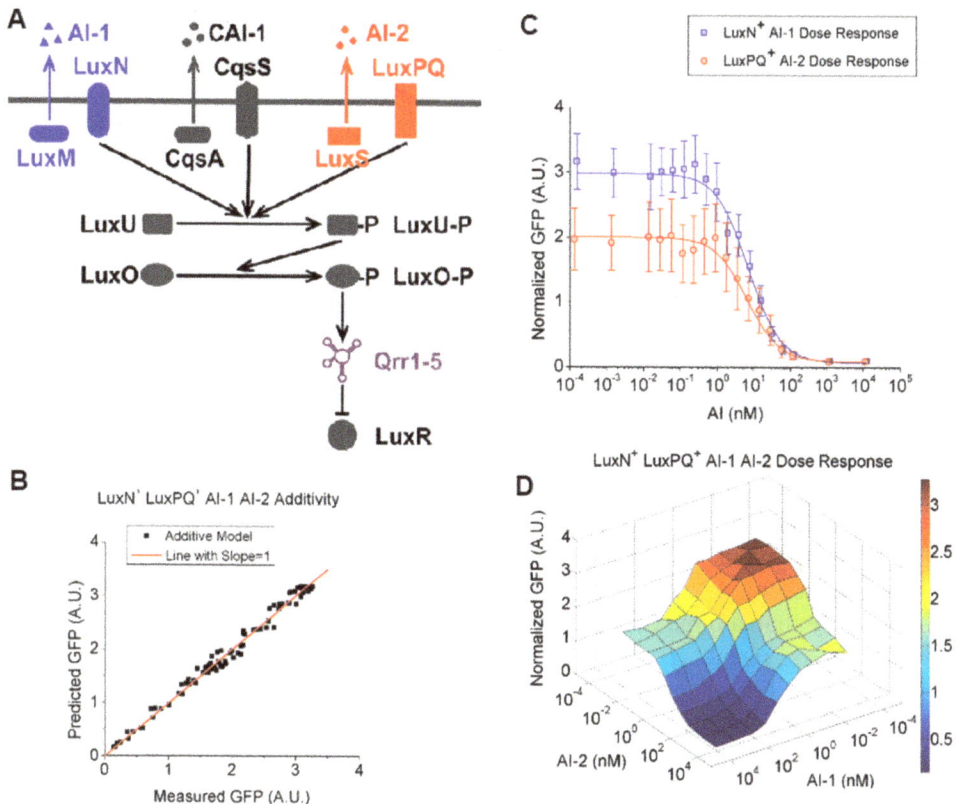

Figure 6.5. Noisy quorum sensing in *V. harveyi*, measured in [12]. (A) *V. harveyi* regulates its bioluminescence using three diffusible signals designated AI1, AI2, and CAI1 which act in parallel. They are detected by the receptors LuxN, CqsS, and LuxPQ, respectively, and trigger the same phosphorelay cascade (involving LuxU and LuxO), ultimately suppressing the master regulator LuxR. (B) The activation of the phosphorelay cascade can be observed in individual cells by inserting in the circuit (at the small regulatory RNAs, Qrr1-5 in (A)) a green fluorescent protein reporter that is activated by LuxO-P. Because the three channels work in parallel, the effects of activating the receptors LuxN and LuxPQ are strictly additive. (C) The green fluorescent protein response from the reporter is reduced as either AI1 or AI2 is supplied, although the response is variable from cell to cell. (D) The combined effect of the two signals is a symmetric step-like activation of the circuit, with both signals contributing equally to the output. Reproduced from [12].

AI2, and CAI1) feeds into the same phosphorelay cascade to regulate one output, known as LuxR[12]. Single-cell observations of this system found that it has generally modest noise, with a coefficient of variation closer to 20%–40%, or less than half that of *V. fischeri* [12]. Nevertheless when the mutual information between each signal input and the output was measured, the result was a modest 0.6–0.8 bits (depending on what assumptions are made about the input distributions). With less than one bit of mutual information, the circuit is not quite able to distinguish between the

---

[12] *V. harveyi* LuxR is not similar to *V. fischeri* LuxR and does not play an analogous role.

presence and absence of one signal [13]. In addition the circuit has the same response to the AI1 and AI2 inputs, so that it cannot even distinguish which of the two signals is present.

Does *V. harveyi* gain more information by using three input signals instead of just one? Again depending on what assumptions are made about the input probabilities of different combinations of signal concentrations, the data indicate that the mutual information between the output and the joint concentration ([AI1],[AI2]) of two signal inputs is perhaps 1.5 bits [13]. The addition of the second signal may enhance the cell's knowledge of its chemical environment, but each signal still delivers no more than one bit of information.

Studies of signal transduction by other biological networks have found similar results. Detailed measurements on some model signal-processing pathways at the single-cell level, with different types of architectures, show that noise at the molecular level limits the mutual information between the signal input and regulated output of such pathways to about one bit [14]. Even if a feedback mechanism is available to reduce the noise, the mutual information is not greatly improved. Although puzzling in some respects, this finding may help to explain the complexity of bacterial communication systems. Any signal processing pathway that needs to provide more than a binary (one bit) response, in order for example to switch the cell between several phenotypes, will require multiple processing channels.

Along these lines it is interesting that many pheromone systems, including *V. fischeri* and *V. harveyi*, employ parallel sensing systems that may interact or crosstalk in an excitatory or inhibitory fashion, and that control a number of phenotypes in addition to bioluminescence. Although there are some speculative ideas about why so many communication circuits have such intricate design, there are no clear answers so far. In *V. fischeri* the presence of C8 HSL can either enhance or inhibit the response to 3OC6 HSL, which leads to a somewhat complex behavior where HSL can either increase or decrease bioluminescence, depending on the combination of signals ([C8 HSL], [3OC6 HSL]) that is present. However, despite this added complexity the mutual information is still less than in *V. harveyi*. Calculating the mutual information between the bioluminescence and the two HSL signals, based on data on the variability of individual cell emission, gives about 0.5 bits of mutual information between state of the HSL inputs and the luminescence output [4]. This may suggest that the two signals provide a more complex function than simple activation, with factors other than concentration (perhaps kinetics or timing) playing a role.

In terms of interpreting *V. fischeri* pheromone sensing, it is interesting that the LuxIR region of the genome shows unusually high variability between strains. There is abundant evolutionary variation in the LuxR receptors of *V. fischeri* strains that inhabit the light organs of different host animals. These receptors have slightly different interactions with the 3OC6 HSL and C8 HSL. LuxR turns out to be a tunable receptor, as illustrated in figure 6.6. While little is known about the sensing behaviors of all the naturally occurring LuxR alleles, synthetic variants of LuxR can

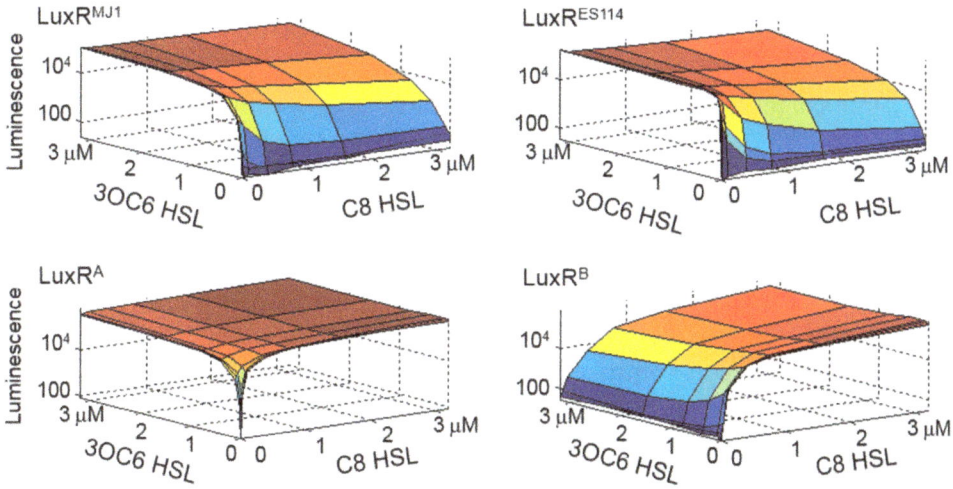

Figure 6.6. Strain-dependent crosstalk between pheromone signals in the LuxIR system. The two signal molecules C8 HSL and 3OC6 HSL interact nonlinearly to produce a complex bioluminescence response in *V. fischeri*, with the nature of the crosstalk depending on the particular bacterial strain. In these data, bacteria that manufactured LuxR (the 3OC6 HSL receptor) but lacked the gene to make AinR (the receptor for C8 HSL) were provided with various combinations of the two signals. The different plots show the bioluminescence response of four different strains, carrying four different alleles (variants) of the LuxR receptor. Depending on the LuxR allele, either signal may enhance or inhibit bioluminescence. Variants LuxR(MJ11) and LuxR (ES114) are derived from *V. fischeri* that inhabit a fish and squid host animal, respectively; LuxR(A) and LuxR (B) are laboratory-engineered variants. Note the logarithmic vertical scale. Adapted from [15].

be designed that vary qualitatively in their response to combinations of the two signals and their stimulation of the *lux* operon. This potential for diverse sensing and activation behavior may imply that the system is capable of triggering different types of phenotypic switches and behaviors depending on the particular needs of the organism in its symbiotic environment.

The idea that one communication system can be tuned to generate different behaviors is embodied in the pathogen *Staphylococcus aureus*, an agent of serious infections in humans. It employs a pheromone-sensing system known as Agr to control its conversion to an aggressive lifestyle as infection develops. The signal is an Agr-inducing peptide, which is detected by a receptor protein AgrC. Curiously, there are four different groups of *S. aureus* strains that produce slightly different versions of the signal peptide. Although each strain can activate its own Agr system using its own peptide, the peptide produced by one group will inhibit the Agr system of a different group. Thus in an environment of different *S. aureus* strains, quorum communication through essentially the same Agr system drives an activating interaction in some strains and an inhibitory interaction in others [16]. These groups in turn are associated with different clinical types of *S. aureus* infections. It is easy to imagine that in a mixed population of *S. aureus* several different overall phenotypic states could be generated or encoded by different combinations of the pheromone concentrations.

6.2 Pheromone triggered transitions of nonlinear systems

Pheromone systems do not just switch a particular gene on or off. More typically they act in combination with other environmental cues to control multiple outputs within the cell. As a result the regulated behavior can be very complex. The pheromone signals drive regulatory pathways that may be noisy and subject to feedback and nonlinearity as well as input from other signaling pathways: chemical communication in a population couples together the individual regulatory networks, each of which is a noisy and somewhat unpredictable dynamical system. This kind of interplay is nicely illustrated by the example of *Bacillus subtilis*, a model organism for so-called cell fate decisions. Its choice between genetic competence and sporulation behavior is controlled by a regulatory pathway that integrates many of the elements discussed in preceding sections: nonlinearity, feedback, bistability, phosphorelays, oscillation, stochasticity, and quorum sensing [17]. The final section of this book gives an overview of how *B. subtilis* assembles these different elements to control cell fate. First, however, we briefly explore the emerging topic of electrical communication in bacteria.

6.3 Electrical communication

Chemical communication in bacteria has been known for decades. A much less understood phenomenon is the use of electrical signals by bacteria. Large scale electrical activity does occur in bacterial colonies, and evidence is accumulating that this activity is a form of communication. An early indication that bacteria do generate electrical currents came from experiments showing coupled electrical activity in different horizontal layers of marine sediments in contact with seawater. (A previous chapter discussed the giant bacteria such as *Beggiatoa* that thrive at these interfaces.) In these environments bacteria at the seawater interface have access to O_2, which they use as an electron acceptor. Bacteria deeper in the sediment are in an oxygen-poor (anoxic) but sulfide-rich environment and must use other molecules such as SO_4^{2-} as electron acceptors. Experiments showed that sulfide-rich marine sediment that was incubated with O_2-rich seawater [18] can rapidly develop three chemically distinct zones, as shown in figure 6.7. An oxygen-rich (oxic) but sulfide-depleted seawater layer is at the top and the sulfide-rich but oxygen-poor (anoxic) layer is at the bottom. These are separated by a centimeter-wide transitional (suboxic) layer. The width of the suboxic layer is found to depend on oxygen availability; the more O_2-rich water is present at the top, the deeper the suboxic region extends, pushing the sulfide-rich layer downward.

This zone movement, on a length scale of about 12–15 mm, occurs too rapidly (about 1 h) to be explained by diffusion of ions. It is more consistent with an electron flow in which the oxidation of sulfide in the sediment is facilitated by an electric current—a flow of electrons up toward the seawater interface, where the electrons can reduce oxygen. One signature of this current is a sharp uptick in pH at the water–sediment interface, where the incoming electrons drive the reaction $O_2 + 4e^- + 4H^+ \rightarrow 2H_2O$ that consumes H^+ ions. These data suggest that microbes

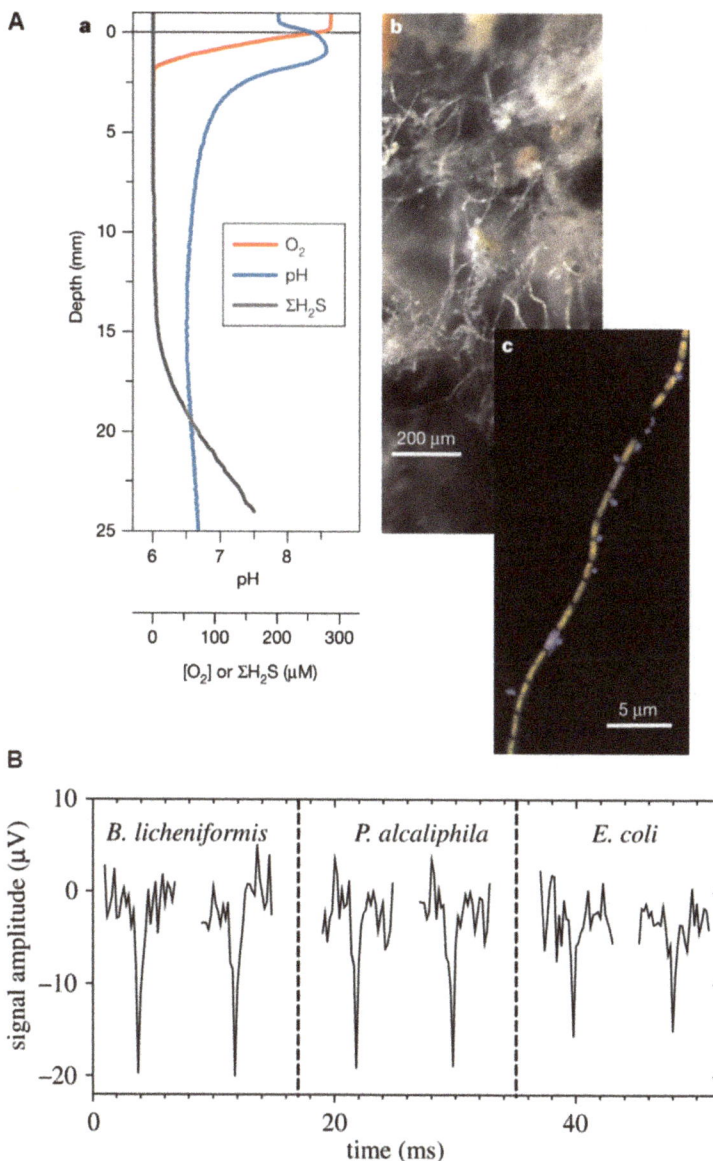

Figure 6.7. Observation of extracellular electrical activity in bacterial colonies. (A) Macroscopic electric current flows through bacterial 'microcables' in a marine sediment. Microbes deep within the sediment have access to electron donors in the form of hydrogen sulfide H_2S and inorganic sulfur S^{2-} but lack access to oxygen as an electron acceptor. Microbes in the upper, oxic zone near the seawater have access to oxygen but lack electron donors. The sulfidic and oxic zones are separated by a 12–15 mm suboxic layer, which is depleted of both sulfide and oxygen (a). *Desulfobulbaceae* bacteria form long multi-cellular filaments (b)–(c) of centimeter length, which conduct electrons up from the sulfidic zone to the oxic zone. One signature of the electric current is the positive spike in pH near the water–sediment interface, where O_2 is reduced to H_2O. Reprinted by permission from Macmillan Publishers Ltd: [19], copyright 2012. (B) Electrical spikes recorded with multi-electrode arrays in biofilms of *Bacillus licheniformis* and *Pseudomonas alcaliphila*, and in planktonic (non-biofilm) *E. coli*. Reproduced from [20] by permission of the Royal Society.

deep within the sediment oxidize sulfide and send the resulting electrons up toward microbes at the surface.

What can conduct such currents? It is possible they could travel through conductive, inorganic minerals. Alternatively, they could travel through networks of bacterial nanowires. Nanowires are a specialized form of pili[13], hair-like protein filaments that extend outward from a bacterium and allow it to make contact with other objects. Some bacteria such as *Geobacter sulfurreducens* use pili to transport electrons. *G. sulfurreducens* oxidizes its carbon source and transfers the electrons along electrically conducting pili to particles of Fe(III), the electron acceptor. A crosslinked system of pili could possibly comprise a nanowire network, providing electrical connectivity over macroscopic distances.

However, the physical basis for the electrical conduction in the sediments is most likely to be the bacteria themselves. Bacteria of the family *Desulfobulbaceae* have been found to form long, entangled multi-cellular filaments within the upper sediment layers [19]. Figure 6.7 shows that while the individual cells are small (3 μm) the filaments can be as long as 1.5 cm, long enough to span the suboxic zone. The conducting medium itself is not known, but direct electrical measurements of the bacterial filaments suggest that current flows along long, uniform ridges that run along the length of the filaments and interconnect the different cells. The conducting wire is evidently located within the periplasmic space[14] so that it is insulated from the external environment by the cell outer membrane [19]. Therefore the electrons delivered by the anaerobically growing cells at the bottom of a filament flow upward, presumably without leaking out of the filament, to be utilized exclusively by kindred cells located thousands of cell lengths distant in the oxic layer. Therefore these electrical 'microcables' divide up the half-tasks of oxidation and reduction, with microbes near the top reducing O_2 and those in the sulfidic layers oxidizing the electron donors [18, 19]

Although the length scale of these currents is impressive, this is not electrical communication unless information is being transmitted. It is also a somewhat indirect observation as actual voltages were not measured. Direct measurements of electrical potentials are not as easily made on bacteria as they are on eukaryotic cells such as neurons, by applying a patch clamp electrode and reading the current. One reason is that the bacterium is only about the same size as a patch clamp pipette tip[15]. Another reason is that there is not much charge to work with. An *E. coli* cell contains a Na^+ concentration near 10 mM and a K^+ concentration of about 30–300 mM, in a volume of about 0.7 μm^3, for about 4×10^6 Na^+ ions and about 10^7–10^8 K^+ ions. An ion channel or an electrical probe that draws a current of just 1 pA would drain this much charge in one or two seconds.

---

[13] Pili is the plural of pilus.
[14] The periplasmic space is the volume between the cytoplasmic (inner) membrane and the outer membrane of a Gram-negative organism.
[15] Actually *E. coli* can be grown into a giant spheroplast, a large single cell that cannot divide, allowing patch clamp studies of its membrane channels.

Nevertheless there is other evidence that bacteria can manipulate electrical potential. The potential difference across the membrane can be detected optically using voltage-sensitive fluorescent probes, which are synthetic dyes or engineered fluorescent proteins that have affinity for the cell membrane. Their position in the membrane subjects them to very large electric fields[16] that arise from the negative potential of the cytoplasm with respect to the exterior. Changes in the electric field modulate the fluorescence of the probe so that electrical changes in the cells can be observed under a fluorescence microscope. In studies using fluorescent protein reporters, individual *E. coli* are observed to undergo a roughly periodic blinking behavior on a time scale of about 1–40 s [21], indicating quick spikes of depolarization of the membrane. Spikes in neighboring cells were uncorrelated however.

The study of this spiking was extended to biofilms, which can be grown on multi-electrode arrays. Here the electrodes have a size (30 μm) and spacing (500 μm) that is much larger than the individual cell, so that each electrode reads a collective, extracellular voltage perturbation that comes from hundreds of individual cells. These data reveal electrical spiking behavior that is spatially and temporally correlated over distances of several electrodes [20], particularly in the biofilm-forming bacteria. Although the significance of these correlations is not understood, biofilm bacteria evidently have a rudimentary, collective electrical behavior, possibly indicative of electrical communication.

Stronger evidence for true, long range electrical communication in biofilms has recently come from imaging of *B. subtilis* biofilms in microfluidic channels [22]. As with other cells, the intracellular potential of *B. subtilis* is negative. Observations using voltage-sensitive dyes show that the biofilm can transiently depolarize; that is, the intracellular potential temporarily becomes less negative. This depolarization can be observed using a cationic fluorescent dye that is attracted to the membranes of hyperpolarized (more negative potential) cells, as in figure 6.8. When a biofilm is imaged in time lapse with such a dye, zones of bright fluorescence (hyperpolarization) are seen to form in the center of the biofilm and then spread outward toward the edges. These patterns indicate coordinated oscillations in the intracellular potential of many cells, occurring with a periodicity of about 3–4 h. They move with a well-defined speed and nearly constant amplitude. This wavelike character implies that they are driven by active propagation, unlike the simple diffusional spreading (with distance-dependent speed and loss of amplitude) that underlies chemical communication in biofilms.

Mechanistically the waves represent a disturbance in the intracellular and extracellular K^+ concentration in the biofilm. The K^+ concentration in the cell interior is much higher than in the surrounding medium. If the extracellular K^+ rises, it can briefly drive an electrical current into the cell, causing the potential inside the cell to rise slightly (depolarization). The cell quickly responds by opening its membrane-bound K^+ channels, allowing ions to flow outward. This outward current

---

[16] In *E. coli* the potential difference across the 4 nm thick membrane is roughly 0.11–0.14 V. This corresponds to a huge electric field, $E \sim 10^7$ V m^{-1}.

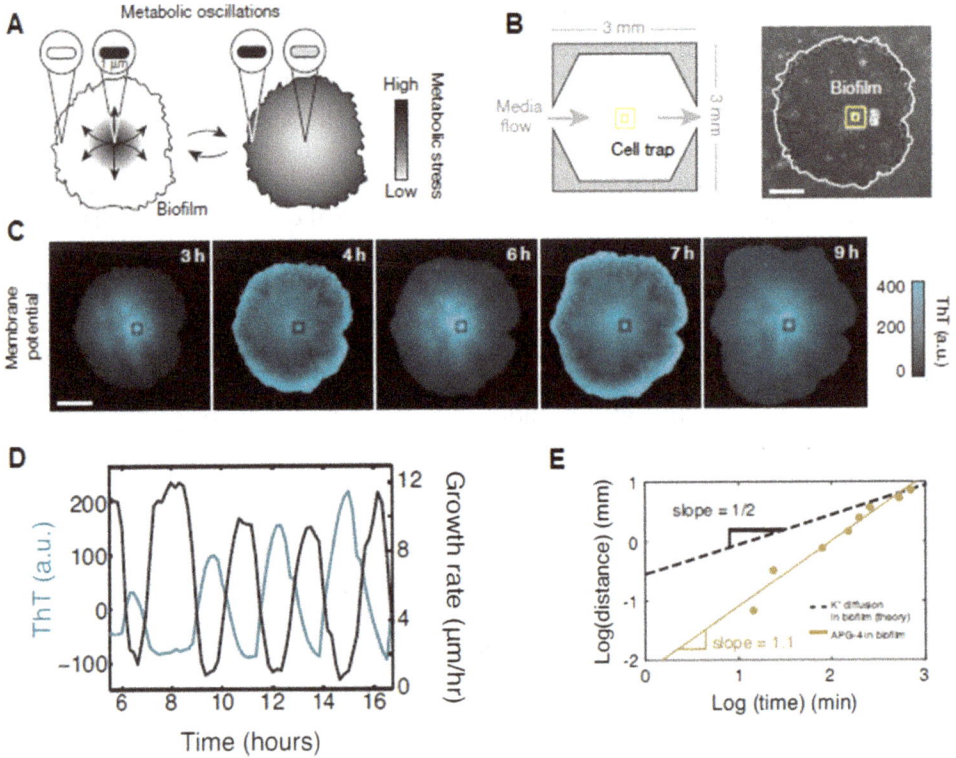

Figure 6.8. Electrical oscillations in biofilms of *B. subtilis* reported in [22]. (A) Biofilms of *B. subtilis* exhibit metabolic oscillations owing to nutrient competition between cells in the interior and periphery. Growth activity alternates between the two groups. (B) Electrical activity in the biofilm can be studied by growing the cells inside a microfluidic flow chamber supplied with thioflavin-T (ThT), a positively charged fluorescent dye that is drawn into the cell membrane in a membrane voltage-dependent fashion. (C) Time lapse imaging of the ThT fluorescence shows waves of hyperpolarization (bright regions) that spread outward from the biofilm center, indicating collective and spatiotemporally synchronized changes in intracellular potential. (D) ThT fluorescence (hyperpolarization) and biofilm growth rate oscillate out of phase, as metabolic slowings are correlated to the intracellular potential. (E) The hyperpolarization travels through the biofilm at near constant speed, characteristic of active wave propagation. A diffusing K^+ signal by contrast would travel a distance that depends on the square root of time. Reprinted by permission from Macmillan Publishers Ltd: [22], copyright 2015.

lowers the intracellular potential (hyperpolarization) although on a much longer time scale, generating the signal that is reported by the fluorescent dye. The outward flux of K^+ raises the local extracellular concentration. This pulse of K^+ is a stimulus that drives adjacent cells to depolarize, and then hyperpolarize. Thus the disturbance propagates through the biofilm, with a speed ~100 μm min^{-1}. Electrical potential becomes spatially and temporally synchronized in regions of biofilm that are separated by large distances, perhaps hundreds of micrometers.

The wavelike behavior is physically quite unlike the diffusive behavior that is typical of pheromone sensing and presumably has a far greater capability to transmit information over long distances. There is reason to believe that it does serve such a purpose. In addition to their electrical waves, *B. subtilis* biofilms can exhibit a metabolic oscillation in which the overall growth rate of the film rises and falls repeatedly on time scales of a few hours [23]. These oscillations appear to reflect competition for nutrients between cells near the center of the biofilm and cells at the periphery, as follows. When the peripheral cells are growing they consume the amino acid glutamate, which is needed by cells in the interior. If growth of the peripheral cells is interrupted, the interior cells can obtain glutamate. This could mean that the electrical depolarization and hyperpolarization of the membrane affects the uptake of nutrients like glutamate. The electrical waves may then provide a mechanism by which interior cells can signal to and modulate the growth of peripheral cells, in order to gain their own access to nutrient. How this plays out at the level of individual cells—especially in terms of the information carrying capacity of these waves—remains to be determined. However, the data so far indicate that electrical communication may prove to be a mechanism of long range metabolic synchronization in biofilm communities.

References

[1] Trushin M V 2004 Light-mediated 'conversation' among microorganisms *Microbiol. Res.* **159** 1–10

[2] Fels D 2009 Cellular communication through light *PLOS ONE* **4** e5086

[3] Williams P, Winzer K, Chan W C and Cámara M 2007 Look who's talking: communication and quorum sensing in the bacterial world *Phil. Trans. R. Soc. Lond.* B **362** 1119

[4] Perez P D, Weiss J T and Hagen S J 2011 Noise and crosstalk in two quorum-sensing inputs of *Vibrio fischeri BMC Systems Biol.* **5** 153

[5] Hense B A, Kuttler C, Mueller J, Rothballer M, Hartmann A and Kreft J-U 2007 Opinion - does efficiency sensing unify diffusion and quorum sensing? *Nature Rev. Microbiology* **5** 230–9

[6] West S A, Winzer K, Gardner A and Diggle S P 2012 Quorum sensing and the confusion about diffusion *Trends Microbiol.* **20** 586–94

[7] Berg H C and Purcell E M 1977 Physics of chemoreception *Biophys. J.* **20** 193–219

[8] Perez P D and Hagen S J 2010 Heterogeneous response to a quorum-sensing signal in the luminescence of individual *Vibrio fischeri PLOS ONE* **5** e15473

[9] Kussell E and Leibler S 2005 Phenotypic diversity, population growth, and information in fluctuating environments *Science* **309** 2075

[10] Shannon C E 1948 A mathematical theory of communication *Bell Syst. Tech. J.* **27** 379–423

[11] Tkacik G, Callan C G Jr and Bialek W 2008 Information flow and optimization in transcriptional regulation *Proc. Natl Acad. Sci. USA* **105** 12265–70

[12] Long T, Tu K C, Wang Y, Mehta P, Ong N P, Bassler B L and Wingreen N S 2009 Quantifying the integration of quorum-sensing signals with single-cell resolution *PLOS Biol.* **7** 640–9

[13] Mehta P, Goyal S, Long T, Bassler B L and Wingreen N S 2009 Information processing and signal integration in bacterial quorum sensing *Mol. Syst. Biol.* **5** 325

[14] Cheong R, Rhee A, Wang C, Nemenman I and Levchenko A 2011 Information transduction capacity of noisy biochemical signaling networks *Science* **334** 354–8

[15] Colton D M, Stabb E V and Hagen S J 2015 Modeling analysis of signal sensitivity and specificity by *Vibrio fischeri* LuxR variants *PLOS ONE* **10** e0126474

[16] Ji G, Beavis R and Novick R P 1997 Bacterial interference caused by autoinducing peptide variants *Science* **276** 2027

[17] Ben-Jacob E, Lu M, Schultz D and Onuchic J N 2014 The physics of bacterial decision making *Front. Cell. Infect. Microbiol.* **4** 154

[18] Nielsen L, Risgaard-Petersen N, Fossing H, Christensen P and Sayama M 2010 Electric currents couple spatially separated biogeochemical processes in marine sediment *Nature* **463** 1071–4

[19] Pfeffer C *et al* 2012 Filamentous bacteria transport electrons over centimetre distances *Nature* **491** 218–21

[20] Masi E, Ciszak M, Santopolo L, Frascella A, Giovannetti L, Marchi E, Viti C and Mancuso S 2014 Electrical spiking in bacterial biofilms *J. R. Soc. Interface* **12** 20141036

[21] Kralj J M, Hochbaum D R, Douglass A D and Cohen A E 2011 Electrical spiking in *Escherichia coli* probed with a fluorescent voltage-indicating protein *Science* **333** 345

[22] Prindle A, Liu J, Asally M, Ly S, Garcia-Ojalvo J and Suel G M 2015 Ion channels enable electrical communication in bacterial communities *Nature* **527** 59–63

[23] Liu J, Prindle A, Humphries J, Gabalda-Sagarra M, Asally M, Lee D, Ly S, Garcia-Ojalvo J and Suel G M 2015 Metabolic co-dependence gives rise to collective oscillations within biofilms *Nature* **523** 550–4

The Physical Microbe
An introduction to noise, control, and communication in the prokaryotic cell
Stephen J Hagen

Chapter 7

Bacillus subtilis competence and sporulation: the final exam

Although we have discussed most of these separately, the design elements of feedback, nonlinearity, and signaling—and the phenomena of bistability, oscillation, and noise—often occur in the same regulatory pathway. In fact, so many of these components may be present in one pathway that the regulatory behavior may be very difficult to interpret. In such cases one can attempt an engineering viewpoint and try to break down the system's function by identifying distinct modules: each module consists of a set of genes or gene products that process information, with modules communicating to generate a behavioral output. A well-studied example is found in *Bacillus subtilis*, a Gram-positive soil bacterium. The regulatory pathway that *B. subtilis* uses to control sporulation and genetic competence behavior is a model of bacterial decision making, integrating nonlinear deterministic dynamics and stochasticity with multiple channels of pheromone sensing. *B. subtilis* combines all of these physical mechanisms in order to control very complex phenotypic switching.

Although it has a normally growing (or so-called vegetative) state, *B. subtilis* can also enter the additional physiological states of sporulation and genetic competence under stress conditions, such as when nutrients are scarce. During sporulation the cell activates a set of genes that induce a form of division in which the mother cell forms an embedded daughter, or endospore. Eventually the mother lyses, or breaks apart, leaving the spore, which can endure harsh physical or chemical conditions such as heat and dessication. Sporulation is a consequential decision for a cell, but it is a rational strategy when severe hardship is inevitable.

doi:10.1088/978-1-6817-4529-9ch7

7-1

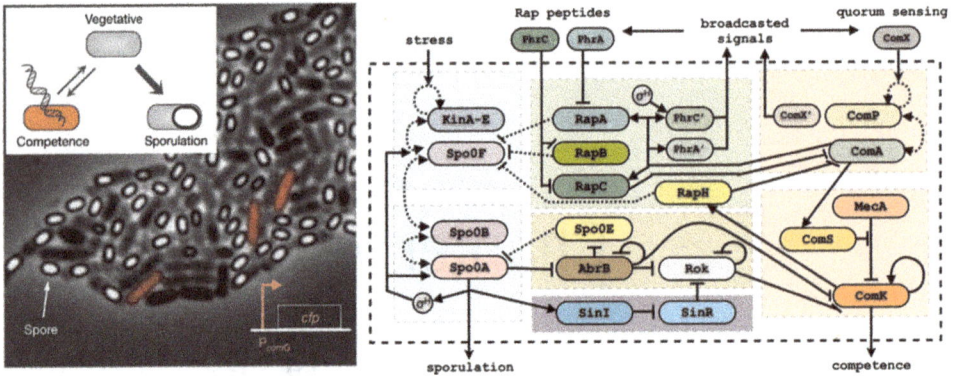

Figure 7.1. The sporulation/competence decision in *B. subtilis*. (Left) Under stress *B. subtilis* activates a regulatory network that drives a decision to enter a transient state of genetic competence or to sporulate. Competent cells, highlighted in red, are highlighted using a cyan fluorescent protein that is under control of the *comG* promoter, which is activated by the master competence regulator ComK. Reprinted by permission from Macmillan Publishers Ltd: [1], copyright 2016. (Right) In this engineering-inspired representation, the competence and sporulation decision is made by a set of interconnected signal transduction and gene regulatory modules. These circuits integrate environmental stress and population density inputs to guide the cell toward sporulation while permitting temporary, stochastic excitation into the competent state. Reprinted from [2] with permission.

Genetic competence[1] by contrast is a transient physiological state in which the cell temporarily has the ability to take up and incorporate DNA from its environment. It requires activation of a large number of genes. However, if other cells in the vicinity have undergone lysis, a competent cell can import some of the genetic material that was released and potentially acquiring new traits. A competent cell can profit from the lysis of its neighbors.

Sustained stress can drive *B. subtilis* into either sporulation or competence, a 'cell fate' decision that each cell makes individually. Adjacent cells may make entirely different decisions, as shown in figure 7.1. If a cell chooses competence it will eventually return to its vegetative state and have another opportunity to become competent or sporulate. The pathway that drives the decision is summarized in figure 7.1. It has a number of recognizable architectural elements and behaviors. By examining some of these elements piecewise one can get a sense of how the organism integrates internal mechanisms with external cues to drive its decision.

7.1 Competence decision by noisy autofeedback

ComK is a transcription factor and the master regulator of competence in *B. subtilis*, affecting the expression of over 100 competence genes. These include the so-called late competence genes whose products bind exogenous DNA, import it into the cell, and recombine and integrate it into the chromosome. When ComK is present at high

---

[1] Competence is short for 'competent for genetic transformation'. A cell in the physiological state of competence is amenable to alteration of its genetic material, a process known as transformation. Competent states occur naturally in at least 80 different species of bacteria.

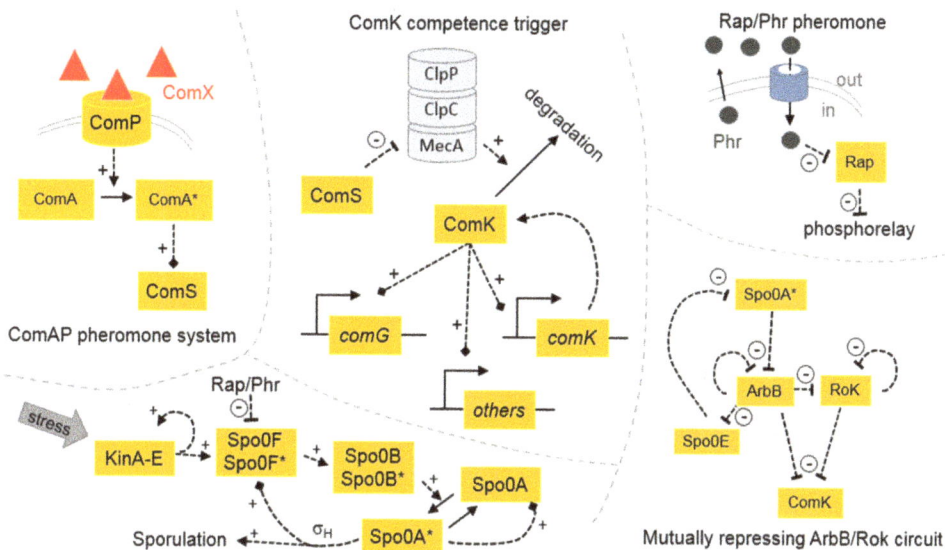

Figure 7.2. Elements of sporulation and competence regulation in *B. subtilis*. The major competence trigger (top center) is the positive feedback loop in ComK, the master competence regulator. ComK acts as a transcriptional activator for numerous genes including those for uptake and incorporation of DNA from the environment. The MecA/ClpP/ClpC protease complex degrades ComK, but ComS inhibits this degradation. ComS in turn is regulated (top left) by pheromone sensing through the ComAP TCSTS. Increased levels of the ComX pheromone act through ComAP to stimulate the competence circuit. The ArbB/Rok circuit (lower right) controls the repression of ComK, with input from SpoOA*. SpoOA*, the phosphorylated form of SpoOA, is the master regulator of sporulation. Its level is controlled by a phosphorelay circuit (lower left) that receives input from the Kin kinases and from the Rap/Phr pheromone system (top right). Phr precursor proteins are cleaved to form peptide pheromones that are exported from the cell. The pheromones, reimported by other cells, inhibit the phosphatase activity of the Rap proteins which inhibit the sporulation phosphorelay.

levels, the cell can enter the competent state. ComK is also its own transcriptional activator. Therefore competence is controlled by a positive feedback loop centered around ComK, as sketched in figure 7.2. However, the cell is not continuously competent because levels of ComK are usually held in check by a degradation mechanism: the protein MecA binds to ComK and targets it for breakdown by a protease[2] system, ClpPC. But this mechanism is itself inhibited by the protein ComS, which offers an alternative target for MecA. ComS competes with ComK for access to MecA, thus preventing ComK degradation. In short, high levels of ComS permit ComK to accumulate, permitting activation of its autofeedback loop and a transition to competence.

The core feedback behavior of ComK is therefore similar to that of figure 3.5 and accordingly the circuit has a similar mathematical model. If S and K are the intracellular concentrations of ComS and ComK, respectively, then a suitable model must include a term for production of ComK and a term for ComK degradation,

---

[2] A protease is an enzyme that breaks down other protein.

taking account also of the competing role of ComS. A greatly simplified model, based on [3], can be written as follows,

$$\frac{\mathrm{d}K}{\mathrm{d}t} = g_0 + g_1\frac{K^m}{K^m + K_0^m} - g_2\frac{K}{K + S + M}, \qquad (7.1)$$

$$\frac{\mathrm{d}S}{\mathrm{d}t} = h_0 - h_2\frac{S}{K + S + M}. \qquad (7.2)$$

The terms g_0 and h_0 describe basal expression of ComK and ComS, respectively, m is a Hill coefficient, g_1 describes the rate of ComK auto-activation, and g_2 and h_2 describe the rates of degradation by MecA/ClpPC. The model as written ignores the different affinities of ComS and ComK for MecA (accounting for both of these with the single parameter M), as well as some apparent repression of ComS by ComK. The model also ignores (of course) the distinction between transcription and translation and various other details. Nevertheless it captures the experimentally observed link between ComK and ComS.

For any fixed ComS concentration S, (7.1) gives $\mathrm{d}K/\mathrm{d}t$ as a function of K. The evolution of K then resembles that of a hypothetical overdamped[3] particle that moves with a trajectory $K(t)$ along an energy surface $U(K)$, which is defined so that $\mathrm{d}K/\mathrm{d}t = -\mathrm{d}U/\mathrm{d}K$, as in the example of figure 3.5. Depending on the parameters, and in particular the concentration of ComS, $U(K)$ may have a double-well shape, indicating a bistable property: low ComK (vegetative) and high ComK (competent) are both potentially stable states. Alternatively, $U(K)$ may have a single-well shape, so that the vegetative state is the only stable state, although the barrier to higher ComK need not be very large. In the single-well case the system would be described as excitable, not truly bistable. These possibilities are shown in figure 7.3.

In a real cell, noisy gene expression leads ComK to fluctuate around its stable states. Sufficient noise can drive ComK over the U barrier from the vegetative to the competent state, and back.

A more complete analysis takes into account the mutual dependence of ComK and ComS. Then the steady states are those for which ComK and ComS are both constant. These are found graphically in figure 7.3 by setting $\mathrm{d}K/\mathrm{d}t = 0$ and $\mathrm{d}S/\mathrm{d}t = 0$ in (7.1) and (7.2) and then finding the intersections of the two resulting S–K curves. The figure also shows how with noisy expression some regions of the (ComK,ComS) plane have higher probability than others, intuitively represented by a bowl-shaped basin of effective potential energy[4]. As above, depending on the parameters the system may or may not have a competent steady state, and that steady state may or may not be stable. Accordingly the competent state may be bistable or excitable.

---

[3] The evolution of $K(t)$ resembles heavily damped motion because it is governed by $\mathrm{d}K/\mathrm{d}t \propto -\mathrm{d}U/\mathrm{d}K$, characteristic of Newtonian motion under strong friction where the acceleration term is insignificant. A lightly damped system would follow $\mathrm{d}^2K/\mathrm{d}t^2 \propto -\mathrm{d}U/\mathrm{d}K$, if such a regulatory mechanism were possible.

[4] The effective potential for a (ComK,ComS) state is the negative logarithm of the probability of finding the cell in that state.

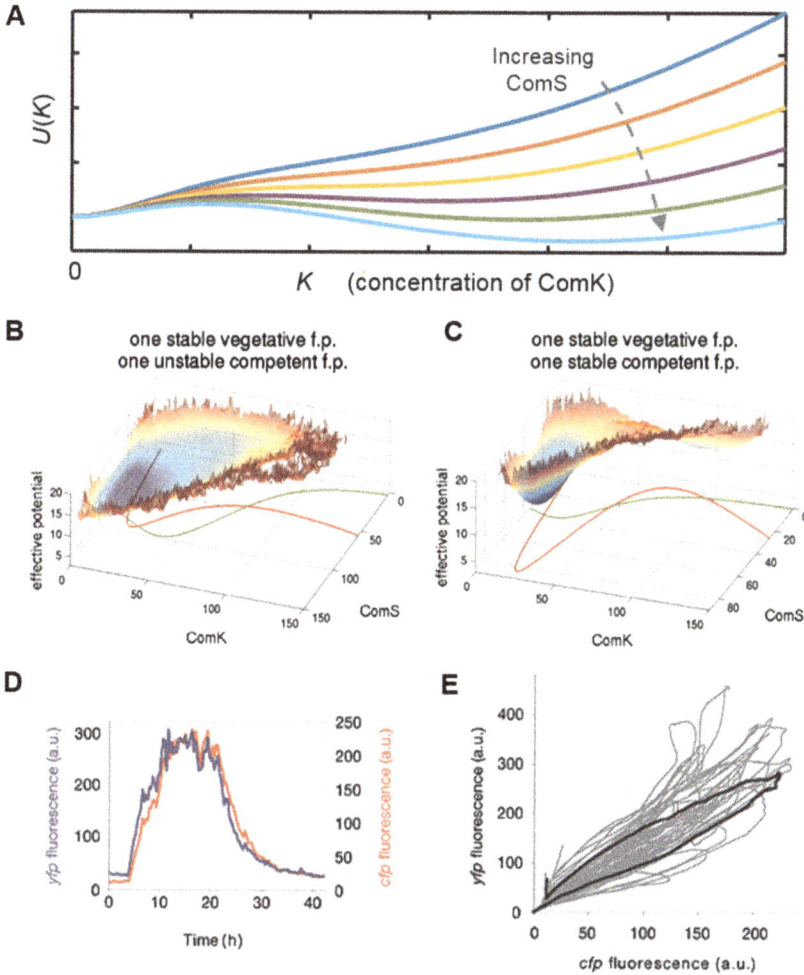

Figure 7.3. Bistable and excitable behavior in the model of (7.1) and (7.2). (A) From (7.1) one can define an effective 'potential energy' U for the evolution of the ComK concentration K, defined so that $dK/dt = -dU/dK$ as in figure 3.5. For some choices of model parameters such as the ComS concentration, the function U may have two minima, indicating that both vegetative (low ComK) and competent (high ComK) states are stable. Or U may have just one minimum, indicating that only transient excursions to high ComK can occur. (B)–(C) The green and red curves correspond to conditions (ComK,ComS) for which $dK/dt = 0$ and $dS/dt = 0$, respectively. The fixed points of the system, at which both ComK and ComS are constant, occur at the intersections of the green and red curves. In the excitable case (B) the vegetative (low ComK) fixed point is a stable steady state but the competent state is an unstable fixed point, while in the bistable case (C) both the vegetative and competent (high ComK) states are stable fixed points. The colored surface is an effective potential derived from the probabilities of different (ComK,ComS) states when stochasticity in gene expression is taken into account. Panels (B) and (C) reprinted by permission from Macmillan Publishers Ltd: [4], copyright 2016. (D) Experimental observation of an individual *B. subtilis* cell undergoing a transient excursion into the competent state. The competence is detected by increased expression of fluorescent protein reporters *yfp* and *cfp* for the ComK and ComG promoters, respectively. (E) An overlay of 37 such individual-cell events, plotted to highlight the correlation between ComK and ComG activity during the event. The correlation and the brief duration of the high ComK/ComG state suggests an excitable (rather than bistable) circuit that passes transiently through its competent state. Panels (D) and (E) reprinted from [1] with permission.

Scenarios of bistability and excitability are shown in figure 7.3. The actual behavior of the competence circuit was observed by tracking the activity of the *comK* promoter in individual cells, using fluorescent gene reporters [1]. This experiment was done for *comK* as well as for *comG*, a late competence gene that is directly activated by ComK. The figure shows that *comK* expression fluctuates over time. Large excursions to higher expression occasionally occur on long time scales, and these are correlated with increased *comG* activity. These data suggest that genetic competence is in fact generated by excitability in a noisy feedback circuit. The cell uses noise and feedback to activate competence intermittently.

7.2 Phosphorelay sensor for sporulation

Sporulation is controlled by a separate pathway based on a phosphorylation cascade. This pathway senses environmental stresses and converts them to an output in the form of phosphorylated Spo0A, denoted spo0A*. Spo0A is the master regulator of sporulation; its concentration and its phosphorylation state control expression of at least 120 different genes. This is primarily because Spo0A* induces synthesis of σ_H, an important regulatory molecule[5] for transcription. The overall sensing circuit has positive transcriptional feedback.

In order for sporulation to occur, the Spo0A* level must attain a threshold value. The phosphorylation cascade that will activate Spo0A* begins with a set of kinases, KinA–KinE, that detect physiological stress such as nutrient depletion and respond by phosphorylating themselves. Once activated these enzymes signal downstream by phosphorylating Spo0F, which transfers the phosphoryl to Spo0B, which passes it to Spo0A to form Spo0A*. In addition Spo0A* is subject to feedback regulation, as its own expression is initiated by σ_H which it induces. Therefore the signaling pathway drives Spo0A* to induce expression of σ_H, which feeds back to stimulate further expression of both Spo0A and Spo0F, and the kinases. As stress conditions persist, the level and activity of Spo0A* rise over time, bringing the cell closer to sporulation. In this way, Spo0A* levels have been compared to a timer that gradually ticks toward sporulation.

7.3 A mutually repressing circuit inhibits competence

The sporulation and competence pathways are linked by a mutually repressing circuit, shown in figure 7.3, that appears to regulate the relative timing of competence and sporulation. The circuit involves Spo0A*, AbrB, and Rok. AbrB and Rok both have autorepressing behavior and they also both repress ComK, suppressing competence. However, their effect on ComK is slightly incoherent, because while Rok and AbrB directly inhibit ComK, ArbB also inhibits Rok. As shown in figure 7.4 this allows for a complex behavior of ComK when ArbB levels change. Meanwhile Spo0A*, which has a stimulating effect on many genes via σ_H, also has a repressing effect on *abrB*. Because AbrB represses Rok, Spo0A* also has

---

[5] σ_H is an example of an alternative sigma factor, a protein that switches on transcription of a large group of related genes by helping the RNA polymerase bind to their promoters.

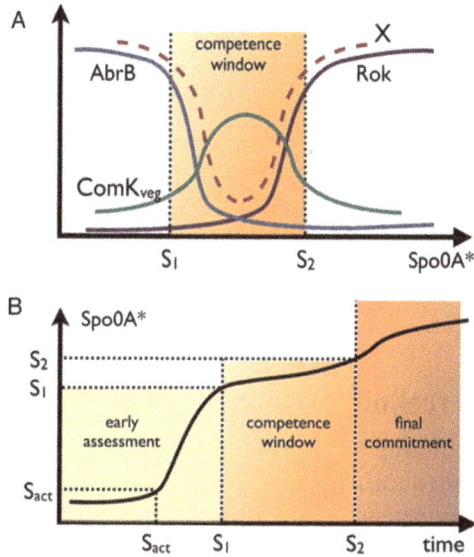

Figure 7.4. Model for Spo0A-AbrB-Rok control of competence timing [2]. AbrB and Rok both repress the competence regulator, ComK, and AbrB also inhibits Rok. (A) Rising concentrations of the sporulation regulator Spo0A* (phosphorylated Spo0A) repress AbrB, allowing ComK levels to rise. However, the loss of AbrB due to increasing Spo0A* permits Rok to rise, quenching ComK. As a result, rising Spo0A* under environmental stress leads to a window of time during which ComK can be expressed at a high level and the cell can enter the competent state. In (B) the feedback from the AbrB-Rok system modulates the accumulation of Spo0A*, causing its concentration to increase at a non-uniform rate. Spo0A* accumulates at different rates through the threshold concentrations S_1 and S_2 where competence becomes possible and then switches off, respectively. At sufficiently high Spo0A* the cell commits to sporulation. Reprinted from [2] with permission.

an indirect activating effect on Rok. In addition AbrB inhibits SpoOE, which dephosphorylates Spo0A*.

Despite the somewhat horrific complexity of these interactions, the essential behavior can be understood by considering the effect of environmental stress that increases Spo0A* levels. At first, rising Spo0A* inhibits AbrB, causing its levels to fall. This alleviates some of the AbrB repression of ComK, permitting higher ComK and a possible transition into the competent state. However, the repression of AbrB derepresses Rok, leading eventually to higher levels of Rok, which repress ComK. Consequently, prolonged repression of AbrB eventually leads to a shutdown of competence.

Therefore the system appears to act with two thresholds, as illustrated in figure 7.4. Under stress conditions, Spo0A* rises to reach the first threshold level at which it inhibits AbrB and competence becomes possible. However, rising Spo0A* levels eventually exert their indirect effect of derepressing RoK, reasserting the ComK repression mechanism and shutting down competence. Competence is permitted only during the time window between the two thresholds.

In addition to manipulating competence, the AbrB/Rok system also manipulates the progress toward sporulation by altering the accumulation rate of Spo0A*. The repression of AbrB allows accumulation of SpoOE, which dephosphorylates

SpoOA*. The progression to sporulation is then slowed while the cell has access to the competence state, as shown in figure 7.4. In this model, the network of mutual repression serves a timing function, delaying sporulation while the cell has the opportunity to enter the competent state.

Finally it is interesting to note that SpoOA* inhibits AbrB, while AbrB inhibits SpoOE, which deactivates SpoOA*. Consequently these three elements form a mutually inhibitory loop that is analogous to the repressilator shown in figure 3.6. Oscillations in this circuit would presumably interact with other environmental cues, generating a very heterogeneous competence and sporulation behavior within a population of cells [2].

7.4 Input from intercellular communication

In addition to these internal dynamics, the circuit receives input from several extracellular pheromone (quorum sensing) signals. As discussed above, ComS plays an important role in competence because it prevents MecA/ClpPC from degrading the master competence regulator, ComK. The level of ComS is controlled by a peptide pheromone, ComX. The cell detects extracellular ComX through a two-component signal transduction system (TCSTS) that consists of (sensor kinase) ComP and (response regulator) ComA. The ComX signal from the cell's environment activates ComA, leading to expression of at least 20 different genes, including ComS. In this way the competence circuitry is modulated in a population-sensitive fashion.

The sporulation cascade also receives quorum-sensing input through the phosphorelay cascade that drives SpoOA*. This is provided by the Rap-Phr system, which comprises a group of phosphatase proteins (denoted Rap), each of whose activity is inhibited by a small peptide (denoted Phr). The Rap proteins act as phosphatases, suppressing the phosphorelay signaling that drives SpoOA-P accumulation. However, the Phr pheromone signals inhibit the activity of their Rap proteins. In this way, they provide population-sensitive input to the action of the stress-sensing kinases KinA–E that drive the cell toward sporulation. This allows the SpoOA phosphorelay of figure 7.2 to weigh two kinds of input: stress information from Kin and population information from Rap/Phr. Signal integration allows a richness that has been revealed in mathematical modeling; not only can sporulation trigger on nutrient deficiency or high population, but the circuit may also be capable of more complex calculations that involve the nutrient availability per cell [5].

The interactions between the different elements of the *B. subtilis* pathway are only summarized here. They can be modeled mathematically at the molecular level [2, 4]. Such work leads to an intuitive physical description in which the cell employs SpoOA* as a sort of internal clock that drives the stochastic switching of the competent state, while also driving the cell progressively toward sporulation, all while probing the environment for population and stress indicators. What is most interesting perhaps is the number of different physical mechanisms and behaviors—intercellular communication, autofeedback, stochasticity, bistability or excitability, phosphorelays and TCSTSs, and even oscillations—that the cell deploys in order to control the overall mechanism. The physical microbe has a powerful toolkit.

References

[1] Suel G M, Garcia-Ojalvo J, Liberman L M and Elowitz M B 2006 An excitable gene regulatory circuit induces transient cellular differentiation *Nature* **440** 545–50

[2] Schultz D, Wolynes P G, Jacob E B and Onuchic J N 2009 Deciding fate in adverse times: sporulation and competence in Bacillus subtilis *Proc. Natl Acad. Sci.* **106** 21027–34 Dec

[3] Ben-Jacob E, Lu M, Schultz D and Onuchic J N 2014 The physics of bacterial decision making *Front. Cell. Infect. Microbiol.* **4** 154

[4] Schultz D, Jacob E B, Onuchic J N and Wolynes P G 2007 Molecular level stochastic model for competence cycles in Bacillus subtilis *Proc. Natl Acad. Sci.* **104** 17582–7

[5] Bischofs I B, Hug J A, Liu A W, Wolf D M and Arkin A P 2009 Complexity in bacterial cell-cell communication: quorum signal integration and subpopulation signaling in the Bacillus subtilis phosphorelay *Proc. Natl Acad. Sci.* **106** 6459–64